Honey

Honey

A Miraculous Product of Nature

Edited by

Rajesh Kumar
Om Prakash Agrawal
Younis Ahmad Hajam

CRC Press
Taylor & Francis Group
Boca Raton New York London

CRC Press is an imprint of the
Taylor & Francis Group, an **informa** business

First edition published 2022
by CRC Press
6000 Broken Sound Parkway NW, Suite 300, Boca Raton, FL 33487-2742
and by CRC Press

2 Park Square, Milton Park, Abingdon, Oxon, OX14 4RN

CRC Press is an imprint of Taylor & Francis Group, LLC

Library of Congress Cataloging-in-Publication Data

Names: Kumar, Rajesh, (Entomologist), editor. | Agrawal, Om Prakash, (Zoologist), editor. | Hajam, Younis Ahmad, editor.
Title: Honey : a miraculous product of nature / Rajesh Kumar, Om Prakash Agrawal, Younis Ahmad Hajam.
Description: First edition. | Boca Raton, FL : CRC Press, 2022. | Includes bibliographical references and index.
Identifiers: LCCN 2021033666 (print) | LCCN 2021033667 (ebook) | ISBN 9781032008257 (hardback) | ISBN 9781032008271 (paperback) | ISBN 9781003175964 (ebook)
Subjects: LCSH: Honey. | Honey--Therapeutic use. | Honey as food.
Classification: LCC RM666.H55 H64 2022 (print) | LCC RM666.H55 (ebook) | DDC 613.2/78--dc23
LC record available at https://lccn.loc.gov/2021033666
LC ebook record available at https://lccn.loc.gov/2021033667

ISBN: 978-1-032-00825-7 (hbk)
ISBN: 978-1-032-00827-1 (pbk)
ISBN: 978-1-003-17596-4 (ebk)

DOI: 10.1201/9781003175964

Typeset in Times
by KnowledgeWorks Global Ltd.

Contents

Preface

With the changing times, new technology has completely taken hold of the health sector, which provides some side effects for free along with the benefits and the product that gives us complete benefits, we have forgotten them, and society is completely discriminate against the benefits of some natural products. But as the technologies further advanced, the side effects of these technologies totally overcome the benefits and then mankind have again slowly started to turn towards herbal, organic and natural products for the solution/treatment of various ailments and illness. From ancient times, nature provides an everlasting priceless service to the mankind in form of natural products. As we enter into the modern world of 21st century, peoples are leaning more towards the nature and natural products, one of them is honey, an important product of honeybees has had a significant place in human's life as medicine from centuries and has wide therapeutic effects. Honey is a complete nutritious organic natural product that provides a complete health benefits for a long time. A lot of research has been done on honey since ancient times. In Ayurveda and Indian Vedas, it has been named as 'Amrut' because the importance of this product starts from the birth of the child and lasts till the last breath of the person. It contains more than 200 substances which increase its quality and these substances give it antibiotic, anti-inflammatory and antibacterial properties. The Indian Buddha community celebrates Madhu Purnima, i.e. the festival of honey, Hindus use it like Panchamruta in the ritual and Christians use it to greet Jewish on the New Year, so that the coming year is full of sweetness. Honey is also known as *Madhu, Madvika, Kshaudra, Saradha, Makshika, Vantha, Varati, Bhrungavantha* and *Pushparasodbhava*. Ancient Egyptians, Assyrians, Chinese, Greeks and Romans also used honey in various rituals, as food as well as in healthcare products. There are many benefits of this wonderful product, which have been reported from time to time through research, and it is also been reported that mixing of honey with other natural components may act as a complement to the existing therapies. Therefore, in the present book, efforts have been made to compile the knowledge of various bee scientists/researchers so that all the aspirants may get benefitted. This book may help to remove various myths about honey, its processing, packaging and trading with special emphasis on medicinal and therapeutic properties of honey.

Editors' Biographies

Rajesh Kumar is currently working as Assistant Professor of Zoology in Department of Biosciences, Himachal Pradesh University, Shimla, Himachal Pradesh (India). He has more than 8 years of experience in teaching and research. His thrust areas of research are Entomology and Environmental Science. During Ph.D., he was awarded JRF by UGC-SAP under the meritorious students in science for 2 years. He was awarded Major Research Project by HIMCOSTE, Shimla (India) on Honeybees and their social impact during findings of which were recognized at national and international levels. He has expertise in beekeeping techniques with the knowledge of sampling and maintaining culture for various insects and animals like earthworms, honeybees and *Lucilia cuprina* flies. He has also completed one self-sustaining project entitled 'Establishment of Vermicompost Unit under Clean Campus Green Campus'. Currently, he is handling 2 incubator projects under HP-STARTUP scheme and 1 major project sponsored by DEST, Government of Himachal Pradesh, India. There are about 25 research papers, 3 books, 5 book chapters, 7 papers in proceedings, 55 abstracts, 9 awards, 12 seminars/conferences and participation in more than 50 national/international conferences in his credit. He has guided 1 Ph.D., 5 MSc students for their dissertation and, currently, guiding 2 Ph.D. and 1 M.Phil. candidate. He is also a member of various scientific agencies and reviewer of peer-reviewed journals.

Om Prakash Agrawal is former Vice Chancellor of Jiwaji University, Gwalior (India). He is former Professor of Zoology in School of Studies in Zoology, Jiwaji University for more than 35 years. During his tenure, he served at many prestigious positions like Director of Indira Gandhi Academy of Environmental Education, Research and Ecoplanning, Head of the Department; Chairman, Board of Studies in Zoology; Chairman, Animal Ethics Committee; Dean Faculty of Life Sciences and Member of Executive Council.

He is a Professor and completed his Ph.D. from Vikram University, Ujjain. He is recipient of CSIR, UGC and ICMR research fellowships. He joined Zoology department, Jiwaji University, Gwalior as Lecturer in 1980, became Reader in 1988, Professor in 1998 and Vice Chancellor in 2006. He has visited Germany for Postdoctoral research during 1983–85 and 1993–94; and other countries like France, Italy, Czechoslovakia, Austria and UK for academic purposes. Prof. Agrawal was awarded prestigious DAAD (Germany); GTZ (Germany) Equipment and Material Grant; major and minor research projects funded by UGC, MP-CST, Ministry of Agriculture and Cooperation; DRDE. He has supervised 27 Ph.D. and 26 M.Phil. students for their research works; attended more than 150 national and international conferences, seminars, symposia, workshops, chaired technical sessions, delivered keynote and plenary lectures; received several awards for research paper and poster presentations. The primary area of research is Entomology (Insect Physiology, Biochemistry and Molecular Biology). He has worked extensively on environmental issues, including environmental conservation and sustainable development, developed expertise in waste management technology, established a novel model Vermicomposting Center for promotion of environment-friendly management of organic waste, developed simple vermicomposting method of household, poultry and animal house waste management; also working on nature-friendly technologies like beekeeping, Spirulina farming, HOME biogas systems. He is currently involved in promotion and consultancy services for solid waste management, superfoods, honey and Spirulina.

Younis Ahmad Hajam holds a doctorate in Zoology from Guru Ghasidas Vishwavidyalaya (A Central University), Bilaspur, Chhattisgarh, India. Currently, he is working as Assistant Professor and Head of Department of Zoology, Career Point University, Hamirpur, Himachal Pradesh, India. He also served as a Junior Research Fellow (JRF) at Department of Zoology, Guru Ghasidas Central University, Bilaspur, Chhattisgarh, India. Dr Younis Ahmad has published 20 research papers in leading national and international journals (such as Life Sciences, Impact factor 3.4) and Toxicology Reports, seven book chapters and two national proceedings. Dr Younis Ahmad has received national award for his contributions in biological science. Dr Younis Ahmad's current research focus is therapeutic management for diabetes/polycystic ovarian cancer, melatonin physiology, antioxidants and natural therapies for other metabolic and endocrine disorders. He is currently guiding five M.Sc. and four Ph.D. students and handling one major research project (Co-PI), three projects under HP Chief Minister Startup scheme. Dr Younis Ahmad has presented his research work in more than 35 national and international conferences; he has attended more than 30 national and international conferences and 5 workshops. He organized various conferences, seminar, workshops and webinar of national and international levels. He has also delivered lectures in international conferences. Dr Younis Ahmad is editorial member of various international journals and is also member of Asian Council of Science editors.

Contributors

Ahsan Ali
Department of Zoology
Panjab University
Chandigarh, India

Natarajan Ashokkumar
Department of Biochemistry and
    Biotechnology
Faculty of Science
Annamalai University
Annamalainagar, India

Bovinder Chand
Department of Zoology
Swami Vivekanand Government
    College
Ghumarwin, India

Ramkesh Dalal
Department of Zoology
Panjab University
Chandigarh, India

Jasvir Singh Dalio
Biology Division
G.G.S.S.S.S. Budhlada
Budhlada, India

Anjli Dhiman
Department of Biosciences
Himachal Pradesh University
Shimla, India

Diksha
Department of Biosciences
Division Zoology
Career Point University
Hamirpur, India

Younis Ahmad Hajam
Department of Biosciences
Division Zoology
Career Point University
Hamirpur, India

Swati Jamwal
Department of Biosciences
Himachal Pradesh University
Shimla, India

Ambothi Kanagalakshimi
Department of Biochemistry and
    Biotechnology
Faculty of Science
Annamalai University
Annamalainagar, India
and
Postgraduate and Research Department
    of Biochemistry
Government Arts College for Women
Krishnagiri, India

Sapna Katnoria
Department of Zoology
Panjab University
Chandigarh, India

Neelima R. Kumar
Department of Zoology
Panjab University
Chandigarh, India

Nitesh Kumar
Department of Biosciences
Himachal Pradesh University
Shimla, India

Rajesh Kumar
Department of Biosciences
Himachal Pradesh University
Shimla, India

Indu Kumari
Department of Biosciences
Arni University
Kangra, India

Shailja Kumari
Department of Biosciences
Division Zoology
Career Point University
Hamirpur, India

Ashish K. Lamiyan
Department of Zoology
Panjab University
Chandigarh, India

Raju Murali
Department of Biochemistry and
    Biotechnology
Faculty of Science
Annamalai University
Annamalainagar, India

Devarajan Raajasubramanian
Department of Botany
Faculty of Science
Annamalai University
Annamalainagar, India
and
Department of Botany
Thiru. A. Govindasamy Government
    Arts College
Tindivanam, India

Vinayagam Ramachandran
Postgraduate and Research Department
    of Biochemistry
Government Arts College for
    Women
Krishnagiri, India

Raksha Rani
Department of Biosciences
Division Zoology
Career Point University
Hamirpur, India

Sunita Saklani
Department of Zoology
Govt Degree College
Sujanpur Tihra, India

Palanisamy Selvaraj
Department of Biochemistry and
    Biotechnology
Faculty of Science
Annamalai University
Annamalainagar, India

Aishwarya Sharma
Department of Biosciences
Division Zoology
Career Point University
Hamirpur, India

Ankush Sharma
Department of Zoology
Sri Sai University
Kangra, India

Bharti Sharma
Department of Biosciences
Division Zoology
Career Point University
Hamirpur, India

Neha Sharma
School of Biological and Environmental
    Sciences
Shoolini University
Solan, India

Preeti Sharma
Department of Biosciences
Division Zoology
Career Point University
Hamirpur, India

Riya Sharma
Department of Biosciences
Division Zoology
Career Point University
Hamirpur, India

Anoop Singh
Department of Translational and
    Regenerative Medicine
Postgraduate Institute of Medical
    Education and Research (PGIMER)
Chandigarh, India

Subramani Srinivasan
Department of Biochemistry and
    Biotechnology
Faculty of Science
Annamalai University
Annamalainagar, India

Veerasamy Vinothkumar
Department of Biochemistry and
    Biotechnology
Faculty of Science
Annamalai University
Annamalainagar, India

1

Evolution of Apiculture, History and Present Scenario

Bovinder Chand
Swami Vivekanand Government College, Ghumarwin, India

Indu Kumari
Arni University, Kangra, India

Rajesh Kumar
Himachal Pradesh University, Shimla, India

CONTENTS

DOI: 10.1201/9781003175964-1

Introduction

Insects are dominant animals on this earth. Usually, insects are considered harmful to humans, but hardly 1% of insect species fall in the pest category. The benefits of insects in maintaining the economy outweigh the injury inflicted. Honeybees are one of the few insects directly beneficial to man. Honeybees are a precious gift to human being given by nature because of their pollination services and cherished products. Beekeeping is an important activity to agriculture, food security, and biodiversity, and it participates in reducing poverty and boosts livelihood in rural areas worldwide. Bees have been the social insects (Wilson, 1971) known to humans since time immemorial. Probably, during the Cretaceous period when the flowering plants evolved on the earth, as other insects, honeybees too co-evolved in nature (Michener and Michener, 1974).

Beekeeping is a scientific area in economic entomology, which comprises the keeping and rearing of bees, their management, honey production, research on bees and bee products with an essential role in agriculture and horticulture. Beekeeping has been practised in many parts of the world, including Asia and Europe, since time immemorial. The first official mentioning of beekeeping dates to about 2400 BC, in official lists of apiarists. Aristotle (384–322 BC) observed the constancy of bees on flowers of specific crop and also mentioned floral fidelity, division of labour and foulbrood disease of honeybees in his famous book *Historia Animalism'*. Romans first time discussed the commercial aspect of beekeeping in 7 BC F. Huber of Geneva performed the first actual scientific study on honeybees in Switzerland in 1792. Description of honey is available in ancient literature of all world cultures. It has been mentioned to be employed as the food of Gods. In Egyptian mummies, honey pots were reported, among other materials and utensils. It is mentioned in the Indian epic Rig Veda (2000–3000 BC) that *Vishnu (Indian God) took the form of a blue bee on a lotus flower.* In the 19th century, the history of beekeeping showed commercial growth when L. L. Langstroth developed a box type wooden hive with several movable frames on which honeybees construct the honeycomb. Also, Comb foundation sheets were designed in 1857 in Germany by Johannes Mehring, and later on in 1896 wax foundation sheets became available for their use in commercial beekeeping. The earliest bees may have arisen in the xeric interior of the paleocontinent Gondwana, which was probably the area of origin for flowering angiosperm plants (Raven and Axeirod, 1974; Deodikar, 1978). This finding indicates the existence of bee fauna for 90 million years. The absence of an adequate fossil record made it impossible to trace the exact ancestral phylogenetic line (Wilson, 1971). However, Evans and Eberhard's (1970) description of one specimen of sphenoid wasp of the Cretaceous period links it to bees' possible ancestor. Thus, the wasp-like ancestors of bees took advantage of the food made available by flowers and began to modify their diet and physical characteristics. The flowers are often nectar and pollen while in the reward: the bees pollinate the flower with the transfer of pollen gamete from one to another flower of the same variety. Available literature suggested that honeybees have functioned as cross-pollinator extensively on a geological time scale (Singh, 2014; Tej *et al.*, 2017). Because of the distinctive pollen-collecting structures and habits, bees are classified in their superfamily, Apoidea (Winston and Michener, 1977; Michener, 1978). The interpretation of fossil specimen suggests the evolutionary trends of social life in Apoidea (Cockerell, 1906; Zeuner and Manning, 1976; Gauld and Bolten, 1988; Ruttner, 1988; Engel, 1999). Based on morphological differences, Maa (1953) divided bee fauna into three distinct genera. Recently, Arias and Sheppard (2005) described the phylogenetic relationships of honeybees (Hymenoptera: Apidae: Apini) by nuclear and mitochondrial DNA sequence data.

The genus *Apis* was most apparently originated in the tropical and subtropical climate, likely India and Southeast Asia (Crane, 1983). The natural geographical distribution of the genus *Apis* shows the most extraordinary diversity in India and the adjacent region with all the species except *Apis mellifera* (Deodikar, 1959, 1978; Michener and Michener, 1974). However, *A. mellifera* presumed originated in the African tropics or subtropics during the Tertiary period, migrating to Western Asia and colder European climates later (Wintson, 1987). Thus, an evolutionary record of various species of bees in the family Apidae, their nesting patterns, comb structure, biometry of individual casts of bees has been used as the tool for classification by various workers (Gerstäcker, 1863; Claridge, 1921; Alpatov, 1929; Stitz and Szebe, 1933; Abushady, 1949; Alber, 1956; Smith, 1961; Dietz, 1962; Ruttner, 1975, 1986, 1988; Seeley and Morse, 1976; Crane, 1992).

Colony Organisation and Age Polytheism in Honeybees

The complex social behaviour must have been evolved in the genus *Apis* during the Miocene era (Ribbands, 1953; Richards, 1953; Michener and Michener, 1974; Seeley and Visscher, 1985). The evolved mechanism of caste polymorphism, i.e. caste differentiation (Beetsma, 1983), and age polytheism, i.e. distribution of labour, seems the key to successful social life in honeybees. Butler (1954) described the method of recognition of the presence of queen by the worker bees. The unique defence mechanism has been seen on the release of alarm pheromone by worker bees (Morse *et al.* 1967; Koeniger *et al.*, 1979). The ability of vision, chemical senses, navigation, dance language for orientation of food source, etc., regulates the overall coordination within the nestmates in bee societies (Lindauer, 1957; von Frisch, 1971; Dyer, 1985; Koeniger, 1986; Doolan, 1995).

The life of worker bees has been estimated to be of about 6 weeks. The first half (3 weeks) is spent for indoor duties (indoor bees), and the second one is assigned for outdoor or foraging (outdoor, foraging, field bees) activities. The worker honeybees show an elaborate system of division of labour (polytheism). The worker's age mainly governs these work tasks; however, according to the colony's need, some modifications may occur. According to the study to be performed by worker bees, their nutritional requirements show due alterations. Newly emerged bees consume a lot of protein (pollen)-rich diet required for the development of their hypo-pharyngeal glands for secretion of royal jelly.

Although honeybees can survive on a pure carbohydrate diet for long periods like most animals, they require proteins, carbohydrates, fats (lipids), vitamins, minerals and water for average growth and development (Loper and Berdel, 1980). All these nutrients must be in the diet, in a definite qualitative and quantitative ratio (Haydak, 1970). These nutritional needs are satisfied by the collection of pollen, nectar/honey, and water. Pollen collected by worker honeybees from a wide range of flowering plants typically meets the dietary requirements for protein, minerals, lipids and vitamins. Nectar collected by honeybees from either floral or extrafloral nectarines of the flower is mainly a source of carbohydrates as 95.0–99.9% of the solids in it are sugars (Table 1.1).

TABLE 1.1

Age-Dependent Polytheism in the Life of Worker Honeybees

S.N.	Age and Designation	Duties	Nutritional Requirements
1.	0–3 days (Sweeper/Cleaner)	Cleaning and polishing of comb cells, warm the brood nest	High-protein diet (bee bread) needed for the development of HPG
2.	3–11 days (Nurse bee)	Nursing brood, queen and drone	Stored honey and pollen/ and bee bread for secretion of royal jelly
3.	12–17 days (Handling bees)	Production of wax, construction and repairing of comb and picking up of nectar from forager bees, its storage, processing into honey	Honey and small quantity of stored bee bread
4.	18–21 days (Guard bees)	Guide the hive entrance from intruders/enemies and ventilation of hive	Nectar and honey
5.	22 days + (Forager bees/field bees)	Forage for nectar, pollen, propolis and water	Nectar

In terms of nutrition, protein is essential for the average growth and development of honeybee tissues, muscles and glands such as hypo-pharyngeal glands (Moritz and Crailsheim, 1987; Schmidt *et al.*, 1995), and it is well known that pollen- and protein-rich diet promotes the development of larvae and ovarian and egg development (Woyke, 1976; Hays, 1984; Wheeler, 1996; Lin and Winston, 1998; Pernal and Currie, 2000; Hoover et al., 2003; Schäfer et al., 2006).

Age-Old Relationship of Man with Bees

A great deal of explanation on bee and their products, especially that of honey, wax and pollen, occurs in various religious books like the Rig Veda (4500–1500 BC), the Atharva Veda and the Grihya-Sutra, the Charaka Samhita, the Upanishads, the Mahabharat, the Ramayana (400 BC), the Quran, the Jewish Talmud and the Bible. The Buddhist literature like Vinaya Pitaka, Abhidhamma Pitaka and Jataka tales also references the use of honey (Wallace, 2018). Foreign travellers like Fahiyan and Hiuen Tsang have mentioned honey as medicine in India. Honey, the most precious edible item obtained by a man from their nests. It is regarded as *Ambrosia*, the food of gods. Honey is regarded as excellent food and medicine by all ancient and modern civilisations. Nutritionally, one tablespoon of honey (about 21 g) contains 17 grams of sugars like fructose, glucose, maltose and sucrose, and it has 64 calories of energy. Chemically, it has approximately 38.2% fructose, 31.3% glucose, 7.1% maltose, 1.3% sucrose, 17.2% water, 1.5% higher sugars like maltodextrin, 0.2% ash and up to 3.2% other nutrients. Other nutritional contents of honey are vitamins like ascorbic acid, pantothenic acid, niacin and riboflavin; along with minerals such as calcium, copper, iron, magnesium, manganese, phosphorus, potassium and zinc, antibiotic-rich inhibin, many amino acids, proteins, phenol antioxidants and other micronutrients (Chepulis, 2008; Saranraj *et al.*, 2016). It has several medicinal properties. For example, it is commonly used to cure cough, cold and sore throat. It has antimicrobial properties and is used to heal wounds, including burn wounds. It helps to improve metabolic activities. It is thought to help in blood pressure management and diabetes also. It is widely used in cosmetic products apart from several food preparations and beverages worldwide (Paul *et al.*, 2007; Cohen *et al.*, 2012; Abuelgasim *et al.*, 2020).

The relationship of man and honeybees is from prehistoric times, as evident from the archaeological rock and wall paintings. They depict honey harvesting by man dating back to the Neolithic age and is still practised by aboriginal societies in Africa, Asia, Australia and South America. Crane (1983, 1992, 1999) has reviewed the ancient literature and bee-related scenes carved on the stones or wall paintings on tombs in ancient Egypt to depict man's relationship with bees. She further enlightened the art of honey hunting originated first in Egypt during 2400–600 BC and later in Europe. The cave painting of a honey seeker depicted in eastern Spain (6000 BC) near Valencia showed the use of a ladder for climbing and a basket-like container to carry the hunted honeycombs. Before learning the art of domestication of bees, man used to rob the bee colonies found in nature, for example, on the trees, on rocks, or in cavities of hollow logs. So, at that time, man cannot be regarded as a 'beekeeper'. Instead, he was a honey collector. From the analysis of these archaeological shreds of evidence, some of which are older than 13,000 years, it can be quickly concluded that honey and beeswax were widely used by all the primary ancient cultures like Babylonians, Chinese, Egyptian, Greek, Mayans, Roman and Vedic.

Association of Bees with the Plants Life

It is the fact that the pollen collected by bees is the 'male seed' of the flower, which fertilises the ovum, was firstly reported by Dobbs (1750). Dobbs (1750) further said that bees gather pollen from only one kind of flower in each visit. Due to this, the character of disastrous cross-fertilisation has been averted. Onwards, several workers reported an intimate role of bees in fertilising the flowers until they collect the nectar and pollen for the nourishment of larval brood and in turns, they work as good pollinator (Betts, 1923; Haydak, 1935; Beeson, 1941; Bhattacharya *et al.*, 1983; Free, 1993). Furthermore, Allen (1998) evaluated the role of bees in the maintenance of richness and biodiversity within the flowering crops. Likewise, in several parts of the world, extensive studies were undertaken to explore the role of bees in the process of cross-pollination and increase in the gross yield from the crops (Narayanan *et al.*, 1961; Gojmerac, 1980; Schemske, 1981; Crane and Walker, 1984; Robinson *et al.*, 1989; Southwick and Southwick, 1992; Bawa, 1990; Abrol, 1992, 1993; Free, 1993; Roubik, 1995; Somerville, 1997; Atwal, 2000; Kremen *et al.*, 2002; Palatty and Shivanna, 2007). Savoor (1996) advocated for managed pollination to achieve the eco-friendly green revolution.

Species of Bees Known for Honey Production

At the moment, total five bee species, namely *A. mellifera*, *A. cerana*, *A. dorsata*, *A. laboriosa* and *A. florae*, are known for honey production. Amongst them, *A. mellifera* and their races are seen in almost all parts of the temperate, tropical and subtropical environment except the poles, while that of the *A cerana, A. dorsata, A. laboriosa* and *A. florae* are found only in Asian countries (Smith, 1961; Crane, 1992; Roy, 1999; Michener, 2000). The bee species, namely *A. cerana* and *A. mellifera*, are cavity-nesting with multi-comb nests where *A. dorsata, A. mellifera* and *A. cerana* (Crane, 1990). Although the economic importance of bees was known to man since the pre civilian age, their scientific rearing methods were tried for 500 years only (Dave, 1954, 1955).

Apis dorsata (Wild Bee/Rock Bee)

This bee species is originated from Asia. They are the largest among all the honeybees and are ferocious. Each colony of *A. dorsata* can yield 30–40 kg of honey on average. Most bee hunters collect honey from these colonies by adopting traditional methods (Sivaram, 2012).

Apis cerana

The species of bee is originated from Asia. There are 12subspecies that are scientifically identified today. Single colony of this species has many combs and less migratory, also easy to domesticate. Absconding is a common phenomenon when the management of the territory is not proper. One colony's annual honey yield will be between 8 and 10 kg in average (Sivaram, 2012).

Apis florea

This bee species is also originally from Asia. It is commonly called a little bee and the smallest in size among honeybees. Five hundred grams of honey may be produced from a healthy colony (Sivaram, 2012).

Apis mellifera

This bee is originated from Europe and Africa, now domesticated in almost all parts of the world. This is the major honey-producing bee species in the world. It shows less absconding and swarming habit. By practising good management practice, it is possible to harvest more than 50 kg of honey from a single productive colony (Sivaram, 2012).

Eco-Biology, Development and Scientific Management of Honey-Yielding Bee Species

Cavity-Nesting Bee Species *Apis cerana* and *Apis mellifera*

In Europe, since the 15th century, a voluminous work on the scientific investigation on biology and behaviour has been carried out worldwide (Sakagami *et al.*, 1980; Wintson, 1987; Crane, 1990; Graham, 1992; Root, 1993; Mishra, 1995). Various fundamental aspects about the life of bees like the 'anatomy and physiology' (Betts, 1923; Snodgrass, 1925; Snodgrass and Erickson, 1992) 'fertilization' (Bishop, 1920), 'musculature' (Morison, 1927), 'life history and development' (Butler, 1975), 'mating behaviour' (Taber and Wendel, 1958), 'pheromones for communication, alarm and other activities' (Butler and Callow, 1968; von Frisch, 1971; Free, 1987; Blum, 1992; Doolan, 1995), 'sting apparatus' (Jayasvati and Wongsiri, 1993; Paliwal and Tembhare, 1998), 'male genitalia' (Simpson, 1960; Koeniger, *et al.*, 1991), 'queen substance' (Butler, 1954), 'role of pheromones in swarm behaviour' (Avitable *et al.*, 1975), 'caste differentiation' (Beetsma, 1983), 'nutritional basis of caste differentiation' (Haydak, 1943; Dietz, 1962), 'sugar and hormonal control' on differentiation of female casts (Asencot and Lansky, 1976), 'colony founding and initial nest design' (Taber and Owens, 1970), 'storing and ripening of honey' (Park, 1925), 'environmental factors controlling the foraging flight' (Abrol, 1988) and 'honey bee pests, predators disease, and bee pathology' (Baily, 1981, Morse and Flottum, 1997) were explored in detail.

Scientific Management of Honeybees for Production of Honey and other Products

Simultaneously, several workers tried to evolve the sustainable bee-friendly mechanism for the commercial production of bee products like honey, wax, pollen, royal jelly, propolis, sting venom, etc., the docile multi-combed cavity-nesting bee species, *A. mellifera* and *A. cerana* (Crane, 1990).

Based on the species-specific characters, the geological distribution, and the nesting habits of cavity-nesting bee species like *A. cerana* and *A. mellifera*, the number of workers contributed to the evolution of a suitable package of practices to rear them under the control of a human.

The credit for the earliest attempts to domesticate honeybees is given to the Egyptians who expectedly started apiculture more than 4500 years ago (Kritsky, 2015, 2017). They used hollow logs, wooden boxes, pottery vessels (terracotta jars), and woven straw baskets or 'skeps' as the artificial beehives, which imitated the natural beehives (Kritsky, 2017). The idea to keep bees in log hives has been reported to come from the fallen tree nested by the cavity-nesting bees. The first mobile beekeeping started in 3000BC in ancient Egypt. The beekeeping was in such a well-developed state in Egypt that it is speculated that to utilise the blooming flowers at different geographical places in the different seasons of the year, there might be the transportation of the bee colonies on the Nile River between the upper and lower corn to get the maximum honey harvest (Yuksel, 2020). In ancient Egypt, beekeeping has been mentioned in Gaius Julius Hyginus, Varro, and Columella (Crane, 1994). A Roman poet, Virgil, wrote beekeeping guides (Crane, 2004). A recipe for mead has been described by Hispanic-Roman naturalist Columella in De re rustic, at about 60 BC (Cook, 1895). The art of beekeeping appears to have reached Greece and Rome from Egypt (Hatjina *et al.*, 2018).

There are also some archaeological pieces of evidence from North America indicating the use of pottery vessels for beekeeping more than 9000 years ago, earlier than Egypt (Crane, 1999). Also, honey hunting appears to be present in Spain, as is depicted from cave drawings believed to be 8000 years old (Traynor, 2008). Beeswax traces from the Middle East from around 7000 BCE have been found

(Roffet-Salque *et al.*, 2015). Chinese archaeologists have found residual samples of honey, rice, and organic compounds kept for use in the fermentation process in the pots belonging to the Bell Beaker Culture of around 2800–1800 BCE (Odinsson, 2010). The depictions from some rock and cave paintings using smoke and fire for honey extraction were already known around 2400–2500 BCE (Bodenheimer, 1960). Preserved honey stored in jars has been found from ancient tombs in Egypt (Kritsky, 2015). In Knossos of ancient Greece, beehives and beekeeping equipment like smoking pots, honey extractors, and other necessary beekeeping instruments have been found. As per the interpretations of Sir Arthur Evans, the beekeeping overseers used to control the highly valued beekeeping industry (Gartziou-Tatti, 2012). At Rehov in the Jordan Valley of Israel, archaeological evidence of beekeeping has been found belonging to the Bronze and Iron Age dating from about 900 BCE (Mazar and Panitz-Cohen, 2010). In Israel by Amihai Mazar, 30 hives have been recovered arranged in orderly rows, which proves the existence of the advanced honey industry 3000 years ago. In his book 'Bee', Claire Preston mentions making honey ale between 300 and 600 BC by the Iron Age people from Scotland, United Kingdom (Preston, 2006). Beeswax was used to make candles and honey was used for making alcoholic mead in the medieval period in Europe. Medieval Europeans were using woven baskets (called skeps) and hollow logs to house their bees (Kritsky, 2017). Beekeeping was also practised in ancient China. The documentation by Fan Li mentions the importance of the quality of wooden bee boxes in the determination of the quality of honey (Pattinson, 2012). In the book 'Golden Rules of Business Success', he has described the art of beekeeping (Zheng *et al.*, 2018). The bees belonging to genera *Apis* and the stingless varieties of bees were also domesticated in some regions of the world like Australia and America (Mason and Mason, 1984). Stingless variety *Melipona beecheii* was domesticated by ancient Maya people (Weaver and Weaver, 1981). *Tetragonula carbonaria*, a stingless bee variety, was used for honey production in Australia (Halcroft *et al.*, 2013).

Till the development of movable comb hives in the 18th century, the early honey collection methods were crude and very cruel. In the process of collecting honey, not only beehives were destroyed, but also the entire colony of the bees was killed or harmed to such a great extent that there were scarce chances of survival and continuity of the colony after honey harvesting (Wildman, 1768). The Europeans started the practice of construction and the use of artificial movable comb hive in the 18th century after understanding bee biology and the social structure of the honeybee colonies. The discovery of beehives with movable frames proved as the revolutionary step in apiculture as it made the harvesting of honey possible without causing much harm or disturbances to the bee colonies. Peter Prokopovich, a Ukrainian beekeeper, is thought to have produced and use movable comb hives on a large scale at around 1806 (Crane, 1992).

Biology of Honeybees

Several scientists and philosophers have contributed to the study of bee biology. Aristotle, the Greek philosopher, and scientist studied and described the honeybees and beekeeping biology. For his studies, he kept bees in primitive hives

(Aygün, 2016). Although he deciphered many bee biology aspects correctly through his observations, he also believed in some ancient myths about the bees. These myths included that bees do not give birth but find their young ones in flowers. He treated the queen bee as the 'king' (Maderspacher, 2007). He believed that honey is distilled from dew or falls magically from the air and is not made by bees at all. According to him, the life span of bees was of 7 years (Aristotle, 1991). Europeans like Swammerdam, René Antoine Ferchault de Réaumur, Charles Bonnet, and François Huber were the pioneers to undertake the scientific study of honeybees in the 18th century. With the use of microscopes, the internal anatomy of honeybees was studied by Swammerdam and Réaumur after the dissections of bee specimens (Cobb, 2002). Bee biology was challenging to understand because of the brutal nature of bees and the presence of a large number of individuals in a colony. Making the studies safer and more accessible, the glass-walled observation hives were first designed by Réaumur (McIndoo, 2017). It made the observation of the bee activities within the pack easier and safer. Through the glass, Réaumur could observe the egg-laying by queens, but he could not see the phenomenon of mating in the bees (Kilani, 1999). It was Francis Huber, a Swiss entomologist, regarded as 'the father of modern honey bee science', for the first time discovered that the queen bee did not mate in the hive but the air (Koeniger, 1986). He was greatly inspired by the study of honeybees from the works of René de Réaumur and Charles Bonnet. He improved the design of glass-walled hives built by Réaumur. These hives could be opened like the leaves of a book allowing the direct observation of hive activities and also inspected individual combs possible in an easier way (Morgenthaler, 1931; Koutchoumoff, 2018).

Even though Huber became blind even before he was twenty, he did a lot of work on honeybees with the help of his wife Marie Aimée Lullin and a very loyal servant François Burns. The bravery, loyalty and devotion of his assistant uncovered the many truths of bee biology as during many investigations, he had to face attacks by the entire bee colony to unearth a single fact. François Burnens conducted careful experiments after receiving instructions from Huber and made observations of bee colonies for him and kept an accurate record of all the statements for more than 20 years. Through his work, the world could know the correct social organisation of the honeybees and the fact that a single queen is the mother of the entire colony (De Candolle, 1832). Huber also supported the Discovery by A.M. Schirach that larvae fed on royal jelly develop into queen bees. He also documented the capability of workers to lay eggs (Johansson, 1955). Destruction of drones after mating from the colony was also described by him and the process of replacement of queen by a new queen. Huber also discovered communication through the antenna by the honeybees among themselves. He was the first to explain the biological history of bee colonies accurately, along with the process of swarming in honeybees. Huber and his co-worker were among the first to study and describe the reproductive organs of honeybees, which included the ovaries and spermatheca of the queen bees and the penis of male drones after dissecting the honeybee specimens under the microscope. Huber published his work as 'Nouvelles Observations Sur Les Abeilles' (New observations on bees) in Geneva in 1792, which documented all the fundamental scientific truths for the biology and ecology of the honeybees (Huber, 1814).

In traditional beehives, the comb construction takes place so that there is usually crosslinking of combs, and it becomes challenging to harvest honey without harming the bees and their brood.

The early methods of honey collection were crude, cruel and unhygienic. At the time of honey harvesting, these methods involved breaking, crushing, and straining of honeycombs. This involved the destruction of the hive and the crushing of various bee stages and these combs, thus leading to destruction and loss of valuable resources and the contamination of the harvested honey. The effect on the beekeeping economy can be easily imagined if the entire colony was destroyed in honey harvesting. It is believed that in Greece, the traditional basket top-bar hives with movable combs, which are now called 'Greek beehives', were used even 3000 years ago, though their oldest reported evidence dates back to 1669 (Harissis *et al.*, 2012). This evidence of the use of removable frame hives in Greece comes from an English clergyman and traveller, George Wheler, who described in the 17th century, the apiculture practices of Greece prevalent (Harissis and Mavrofridis, 2012). These hives may be regarded as the forerunners of the modern movable frame hives.

Before the development of removable frame beehives, the bees were treated as an annual crop. At the time of honey harvesting, beekeepers used to kill the bees to access their honey. These beehives were closed structure with just single openings for entry or exit to and from the hive. For accessing honey, there was no way other than killing the bees and destroying the hive in the honey harvesting process. For killing bees at the time of harvesting, different methods were used. Mainly the smoke and fire were used to suffocate and kill the bees. In Europe, a piece of burning sulphur was kept at the hive entrance until the bees were suffocated and killed.

The actual revolution in beekeeping took place in the 18th and 19th centuries after the shift from the use of destructive beekeeping skeps and log hives to the hives with removable frames after contributing to the revolutionary work of an Englishman, Thomas Wildman. The publication of 'A Treatise on the Management of Bees' in 1770 by Thomas Wildman can be considered the turning point in the history of beekeeping. His work was based upon the earlier studies and findings in the beekeeping biology done by Swammerdam, Maraldi, and de Réaumur. The result of Réaumur's on the natural history of bees was particularly reviewed by him thoroughly. Earlier efforts by the others in hive design were also taken into account by him, especially the work from Brittany dating from the 1750s, due to Comte de la Bourdonnaye.

Wildman described the design of a new hive that could make honey harvesting possible without the inhuman killing of bees or destroying the pack and so, making it possible to maintain the continuity of the colony. Wildman's style hive involved a skep with an open top and a woven lid that can be removed. This hive used seven frames hung from the top, which is utilised by the bees to build separate combs rather than forming interlinked combs inside the pack. It made it easier to harvest the honey without destroying the hive and harming the bees. It made a massive difference to the way people used to keep bees and harvest honey before this publication. He also gave an idea of the multistorey configuration of the hives, which can be regarded as a forerunner to supers of modern beehives. He gave an idea of adding multiple skeps below at the proper time so the bees could move to a new one, leaving the old upper ones filled with honey relatively free of brood and bees (Wildman, 1768). In this way, bees could be preserved at the time of harvest. The modern movable comb hives are based upon another idea of incorporating the 'sliding frames' to be used by the bees to build their comb by Wildman. The first practical movable comb beehive was designed in 1838 by a Polish beekeeper Johann Dzierzon, making it possible to manipulate the hive combs without hive destruction. There are many contributions of Dzierzon in beekeeping science other than the Discovery of a perfect sliding frame hive. All modern beehives are based upon his design. In 1845, Dzierzon described the development of drones through parthenogenesis and the story of the queen and workers from the fertilised eggs. He also told the role of diet variations in determining the future roles of the female bees (Everett Mendelsohn and Allen, 2002). He is regarded as the father of modern apiology and apiculture (Crane, 1999).

The landmark invention of Langstroth (1853) on 'bee space' by using top-bar movable frames enabled movement of bees inside the hive, i.e. between the parallel combs. The second significant development of the mid-19th century was the advent of comb foundation (Johansson and Johansson, 1967). This comprises a sheet of pure beeswax embossed on both sides with the base and beginning of the cell wall of the comb of honeybee. The comb foundation sheet is fixed in the frames and provided to the bees as the starter comb. Simultaneously, the technique of extraction of honey from honeycombs using centrifugal force (without squeezing) was invented by Major F. Hruschka in Austria in 1865 (Crane, 1990, 1992). These three inventions proved to be the foundation of modern beekeeping and made bees rearing within reach of commons.

In the 19th century, Lorenzo Lorraine Langstroth, a native of Philadelphia and 'the father of American Apiculture', made many contributions to industrialised beekeeping practices. He published a classic book, 'The Hive and Honey-bee', in 1853 (Langstroth, 1857). He accurately calculated the bee space as less than 9 mm but greater than 6 mm. The Discovery of bee space was based upon the fact that bees have a particular spatial preference on how their combs are spaced.

This Discovery can be considered another central turning point in the evolution of apiculture and the field of designing beehives. In 1851, he designed a beehive based upon his experimental work and also from the inputs of the earlier studies made by Dzierzon, Huber and others by keeping the correct bee space among the combs and also between the combs and sidewalls of the hive (Ren, 2020; Attridge and Bagster, 2011). Although Langstroth is often credited with discovering the 'bee space', but Dzierżon had already implemented it to some extent (Attridge and Bagster, 2011). But the Langstroth design was so perfect that it is even followed by the modern beekeepers, as his calculations of bee space was very accurate. The concept of proper bee space removed the possibility of cross-comb formation. He used specialised removable plates, which can be individually removed easily for inspection of the hive or for honey extraction without causing any harm to the bees. It made the addition of new frames or removed the old structures from the pack easier at the time of need (Langstroth, 1852; Ebert, 2009; Abu and Sahile, 2011; Ren, 2020). It also made the translocation of the hives from one place to another place much easier as the chances of damage to the hive got minimised due to lesser possibility of collisions between combs due to jerks etc. The development of a movable comb hive and the rediscovery of bee space were published by him in the book 'A Practical Treatise on the Hive and Honey-Bee' in 1878 (Langstroth, 1857). It revolutionised commercial apiculture on a massive scale in both Europe and the United States. After that, the new tools, techniques and methods for honey harvesting, especially the honey extractor's discovery and use, further minimised the harm to the combs during the harvesting process. The bees, in this way, can get intact combs for refilling without wasting time in the construction of the new combs.

English inventor and beekeeper William Broughton Carr invented in 1890 the double-skinned wooden WBC beehive (named after the inventor), having a pitched roof, short legs, and multistoreys with sloping sides, giving it a pagoda-like appearance thought to provide better insulation for the bees from hot and cold weather. It is presently used only by a minority of amateur beekeepers in the United Kingdom (Malaka and Fasasi, 2005; Fasasi, 2016).

Langstroth's design of beehive made a global impact in the field of apiculture. Based upon this design, several other types of hive designs were conceptualised in different parts of the world. Despite some variations, all these hives are fundamentally similar as these consist of a floor, brood box, honey super, crownboard and roof. They all are square or rectangular and use movable wooden frames (Kasangaki *et al.*, 2014). Though modern hives are formed from wood usually, recently polystyrene-based packs are also available and becoming famous. The polystyrene-based hives are supposed to be better than traditional wooden beehives in many aspects. They provide better insulation from cold and are free from rot and woodworms.

As various factors related to beekeeping vary from one region to another, there are variations in hive designs from one area to another. Each part has its classical hive design. These factors include different bee flora, different bee varieties, differences in bee biology and behaviour. Langstroth and Dadant designs are widely used in the United States and in some parts of Europe (Riedel, 1967; Biswas, 2020; Requier *et al.*, 2020). Dadant hive with larger frames suits some regions and climates. Modifications to Dadant design later gave rise to 'Dadant-Blatt', 'Modified Dadant', or 'Langstroth Jumbo' hive designs.

In Scotland, the smaller Smith hive design is popular. In France, the De-Layens trough-hive, a horizontal hive with extra-deep frames, all frames the same size, is widespread, which provides large combs; for instance, spring build-up helps the bees in overwintering efficiently. Many other countries have their national hive designs. In the United Kingdom, a British National hive is popular in both commercial and amateur beekeepers. Due to its modular design, it is easy to move this hive from one place to another.

An essential addition to the modern hives to the Langstroth beehive is the insertion of queen bee excluders so that the queen doesn't reach the super upper compartments, thus preventing the laying of the eggs there and avoiding the contamination of the honey during harvesting (Doolittle, 1915).

Bee colonies are always prone to infestation by the pests like mites, and to overcome it, the wire mesh and removable trays are often used in place of hive floors. In 2015, the father and son duo invented a new hive design, Stuart Anderson and Cedar Anderson, called a honey flow hive. It involves the minimum disruption to the bees when the honey is harvested. With a honey flow hive, there remains no need to open the hive box or take out the frames. It also makes honey extraction possible without expensive centrifuge equipment (Allan *et al.*, 2015).

The task of designing various bee equipment like 'wooden boxes' for hiving the cavity-nesting bee species (named 'Langstroth hive' for the *A. mellifera* and 'Newton hive' for the *A. cerana*), movable frames of wood for accommodation of the combs, smoker for pacifying the stinging attitude of bees, bee-suit for the protection against stings, hand gloves, hive tool, brush to remove bees from the combs, swarm capturing net called bee veil, water pot, ant arrester, centrifugal honey extractor, vessels for storage of honey and modern processing apparatus, pollen traps, queen cages, feeders, queen gate, queen excluder etc. has been undertaken in many parts of the world as per the geographical variation of bee species (Crane, 1990). Simultaneously, the work on other essential aspects like 'bee pests, predators and diseases' (Morse and Flottum, 1997) 'an art of artificial queen rearing and package bees' (Morse, 1979, Laidlaw, 1992), etc. were also adequately studied to supplement the science of beekeeping.

This is called modern beekeeping science and it is extensively adopted world over (Smith, 1960; White, 1967; Singh, 1975; Wedmore, 1979; Rawat, 1982; Verma, 1991; Graham, 1992; Crane, 1992; Shende and Phadke, 1993; Chahal *et al.*, 1995; Mishra and Sharma, 1997; Suryanarayana and Subba Rao, 1997; Thomas *et al.*, 2002). It proves to be an essential input in agriculture, rural employment, human nutrition and economic development.

Open-air Nesting Asiatic Primitive Bee Species
Apis florea, Apis dorsata, and *Apis laboriosa*

Unfortunately, the concept of modern beekeeping does not apply to the rearing of the Asiatic primitive bee species, namely *Apis Florea, A. dorsata* and *A. laboriosa*. On the one hand, many workers were contributed decently to developing suitable technologies to harvest the bee products from cavity-nesting bee species with eco-friendly inputs, but the Asiatic primitive wild bee-nesting bee species remained obscure (Crane and IBRA, 2001). It seems that due to their diverged nesting patterns and some *A. dorsata* in particular. As a result, today, a unified approach and package of practices of their eco-friendly handling have not yet been worked out properly by the scientific community to serve the purpose of honey and wax production in a scientific manner.

Main Diseases of Honeybees

Honeybees are susceptible to various diseases, some of which are very contagious and can be easily transmitted (Table 1.2). The occurrence of diseases in honeybees depends upon three factors:

1. The genetic heritage of queen bee
2. Pathogens
3. Environmental factors

TABLE 1.2

Main Diseases of Honeybees Classified by Nature of Pathogen

Disease	Causative Agent	Type
Acariasis	*Acarapis woodi*	Parasitic
Aethinosis	*Aethina tumida (small hive beetle)*	Parasitic
Tropilaelapsosis	*Tropilaelaps* spp.	Parasitic
Varroosis	*Varroa destructor*	Parasitic
American foulbrood	*Paenibacillus larvae*	Bacterial
European foulbrood	*Melissococcus pluton*	Bacterial
Chalkbrood	*Ascosphaera apis*	Fungal
Nosemosis	*Nosema apis – Nosema ceranae*	Fungal
Amebiasis	*Malpighamoeba mellificae*	Fungal
Acute bee paralysis virus (ABPV)	*Dicistroviridae*	Viral
Black queen cell virus (BQCV)	*Dicistroviridae*	Viral
Chronic bee paralysis virus (CBPV)	*Cripaviridae*	Viral
Deformed wing virus (DWV)	*Iflaviridae*	Viral
Kakugo virus	*Iflaviridae*	Viral
Kashmir bee virus (KBV)	*Dicistroviridae*	Viral
Sacbrood virus (SBV)	*Virus picorna-like*	Viral

Source: Food and Agriculture Organization of the United Nations, Rome, 2020 ISSN 2708-115X.

Present Status of Apiculture

The economics of honeybees will never be complete without taking into account their role in plant pollination. According to an estimate, 70% of about 100 cultivated plant species, which are the source of 90% of food to the entire world, are pollinated by bees and other native insects. The bees pollinate 80% of natural plants and 84% of commercial crops of Europe, and in terms of money, this corresponds to 22,000 million euros annually. Their contribution to pollination for the rest of the world is estimated worth 265,000 million dollars (Ranz, 2020).

Demand for honey and honey-based products are increasing day by day due to the awareness about the benefits of honey in maintaining a healthy lifestyle. At present, honey is being used as a preservative and an alternative to artificial sweeteners in various processed food products, so the demand from food processing industries boosts the growth of the apiculture market at global levels. Honey has been in use in several traditional medicines, especially in Ayurvedic, Egyptians, Assyrians, Chinese, Greek and Roman, and all other traditional systems of health and medicine for ages. Honey and other hive products are an inseparable part of several cosmetics and pharmaceutical products apart from food and beverages. Honey is thought to boost to enhance immune response. Several clinical trials of honey on COVID-19 patients are currently undergoing (Hossain *et al.*, 2020; Tantawy, 2020). The cosmetics industry in modern times is focussing on products based upon natural and organic ingredients of which honey shares a significantly large part, which is also contributing to the growth of the global apiculture market. Demand for honey increased in 2020 due to the COVID-19 pandemic due to its excellent medicinal properties. In the year 2020, the global honey market size was estimated worth USD 9.21 billion, and it has the potential to achieve a CAGR of 4.3%–8.2% within the period 2020–2028. Major production centres of honey are North and South America, Asia, and Europe. China, at present, is the chief exporter of honey in Europe and North America in the past few years (https://www. grandviewresearch.com; https://www.businesswire).

The largest consumers of honey and its by-products are at present Europe and North America, while rapid growth has been projected in the Asia-Pacific regions. With a 42% market share, the Asia-Pacific was the leader till 2018, followed by Europe. China alone was the largest producer of honey with 230,000 tons of honey production, and Europe claiming the second spot (https://www.businesswire).

China was the leading exporter of honey in 2019 with honey exports worth US$235.3 million (11.8% of exported natural honey), followed by other major exporters New Zealand: $228.8 million (11.5%), Argentina: $146.7 million (7.4%), Germany: $131.5 million (6.6%), Ukraine: $113.3 million (5.7%), India: $99.6 million (5%), Spain: $92.1 million (4.6%), Hungary: $82.5 million (4.1%), Brazil: $67.9 million (3.4%), Belgium: $64.1 million (3.2%), Vietnam: $57.4 million (2.9%), Mexico: $55.7 million (2.8%), Romania: $44.5 million (2.2%), Poland: $43.2 million (2.2%), and Canada: $41.3 million (2.1%). 75.5% of global natural honey export market was shared by these 15 countries in 2019 by value (http://www.worldstopexports.com).

REFERENCES

Abrol, D.P. 1988. Effect of climatic factors on pollination activity of alfalfa-pollinating sub-tropical bees *Megachile nana Bingh* and *Megachile falvipes Spinola* (Hymenoptera: Megachilidae). *Acta Ecol.*, *9*, 371–377.

Abrol, D.P. 1992. *Bee pollinators and host plant relationship. Sci. Cult. (India)*, *58*, 228–230.

Abrol, D.P. 1993. *Insect pollination and crop production in Jammu and Kashmir. Curr. Sci.*, *65*(3), 265–269.

Abu, T. and Sahile, G. 2011. On-farm evaluation of bee space of Langstroth beehive. *Livest. Res. Rural. Dev.*, *23*(10), 207.

Abuelgasim, H., Albury, C. and Lee, J. 2021. Effectiveness of honey for symptomatic relief in upper respiratory tract infections: a systematic review and meta-analysis. *BMJ Evid. Based Med. 26*(2), 57–64.

Abushady, A.Z. 1949. Races of bees. In: *The hive and the honey bee* (R. Grout, ed.), Dadant and Sons, Hamilton, 11–20.

Alber, M.A. 1956. Multiple mating. *Brit. Bee J.*, *83*, 134–135.

Allan, S., Anderson, C., Anderson, S., Bahou, A., Blair, S., Brown, H. and Smith, B. 2015. Australian Story: Going with the Flow. https://www.abc.net.au/austory/the-flow-on-effect/7834664.

Allen, W.G. 1998. The potential consequences of pollinator decline on the conservation of biodiversity and stability of food crops yields. *Conserv. Biol.*, *12*, 8–17.

Alpatov, W.W. 1929. Biometrical studies on variation and races of the honey bee (*Apis mellifera* L.).*Q. Rev. Biol.*, *4*(1), 1–58.

Arias, M.C. and Sheppard, W.S. 2005. Phylogenetic relationships of honey bees (Hymenoptera: Apinae: Apini) inferred from nuclear and mitochondrial DNA sequence data. *Mol. Phylogenetics Evol.*, *37*(1), 25–35.

Aristotle, A. 1991. *History of animals* (Vol. 3). Harvard University Press, Cambridge, MA.

Asencot, M. and Lansky, Y. 1976. The effect of sugars and juvenile on the differentiation of female honey bee *(Apis mellifera* L.) larvae to queens. *Life Sci., 18,* 693–700.

Attridge, A. and Bagster Jr, S. 2011. Adey, Margaret. "Nairobi Conference on Tropical Apiculture." Bee World *66*(2), 54–58. Allen, Mrs. Armstrong. "Dignity of Beekeeping." Victorian Bee Journal 1, no. 5 (1919): 10. Amdam, Gro. "Overview of Research in Beekeeping." Interview, August 16, 2008. *Beeconomy: What Women and Bees Can Teach Us about Local Trade and the Global Market,* 337.

Atwal, A.S. 2000. *Essentials of beekeeping and pollination.* Kalyani Publishers, Ludhiana, India, 395.

Avitable, J.L., Morse, R.A. and Boch, R. 1975. Swarming honey bees guided by phero-mones. *Ann. Entomol. Soc. Am., 68,* 1079–1082.

Aygün, Ö. 2016. *The middle included: Logos in Aristotle.* Northwestern University Press.

Baily, L. 1981. *Honeybee pathology.* Academic Press, London.

Bawa, K.S. 1990. Plant pollinator interactions in tropical rainforests. *Annu. Rev. Ecol. Syst., 21,* 399–422.

Beeson, C.F.C. 1941. *Ecology and control of the forest insects of India and the neighbour-ing countries.* Dehradun, India, The Author, 767.

Beetsma, J. 1983. Measures to control honeybee diseases in the Netherlands. In: *Varroa Jacobsoni Oud. affecting honey bees: Present status and needs: proceedings of a meeting of the EC Experts' Group,* Wageningen, 7–9 February 1983.

Betts, A.D. 1923. Practical bee anatomy. In: *The Apis club,* Benson, Oxon, England, 88. The Apis Club, 1923.

Betts, A.D. 1923. The identification of pollen sources. *Bee World, 5,* 43–45.

Bhattacharya, K., Gupta, S., Ganguly, P. and Chandra, S. 1983. Analysis of the pollen load from a honey sample from Salt Lake City, Calcutta (India). *Sci. Cult. (India) AGRIS Rec., 49*(7), 222–224.

Bishop, G.H. 1920. Fertilization in the honeybee. *J. Exp. Zool., 31,* 225–265.

Biswas, S. 2020. *Intelligent beehive status monitoring in noisy environment* (Master's thesis).

Blum, M.S. 1992. *Honey bee pheromones.* In: *The hive and the honey Bee* (Graham, J.M. ed), Dadant and Sons, Inc., Hamilton, IL, 374–394.

Bodenheimer, F.S. 1960. *Animal and Man in Bible Lands: Supplement* (Vol. 10). Brill Archive.

Butler, C.G. 1954. The importance of 'queen substance' in the life of a honeybee colony. *Bee World, 35*(9), 169–176.

Butler, C.G. and Callow, R.K. 1968. Pheromones of the honeybee *(Apis mellifera* L.): the "the inhibitory scent" of the queen. *Proc. Roy. Entoml. Soc. London (B), 43,* 62–65.

Butler, C.G., 1975. *The honey bee colony-life history.* In: *The hive and the honey bee.* Dadant and Sons, Inc. Hamilton, IL, 39–74.

Chahal, B.S., Brar, H.S., Gatoria, G.S. and Jhajj, H.S. 1995. *Italian honeybees and their management.* Punjab Agricultural University, Ludhiana, 116.

Chepulis, L. 2008. *Healing honey: A natural remedy for better health and wellness.* Universal-Publishers Irvine, California.

Claridge, F.M. 1921. Wintering bees. *Bee World, 3*(7), 185–187.

Cobb, M. 2002. Biographical article Jan Swammerdam on social insects: a view from the seventeenth century. *Insectessociaux, 49*(1), 92–97.

Cockerell, T.D.A. 1906. The North American bees of the family Anthophoridae. *Trans. Am. Entomol. Soc. (1890), 32*(1), 63–116.

Cohen, H.A., Rozen, J., Kristal, H., Laks, Y., Berkovitch, M., Uziel, Y. and Efrat, H. 2012. Effect of honey on nocturnal cough and sleep quality: a double-blind, randomized, placebo-controlled study. *Pediatrics, 130*(3), 465–471.

Cook, A.B. 1895. The bee in Greek mythology. *J. Hellenic Stud.*, *15*, 1–24.

Crane, E. 1983. *The archeology of beekeeping*. Duckworth, London, 360, 270.

Crane, E. 1990. Bees and beekeeping: Science. In: *Practice and world resources.* Comstock Publishing Associates, Cornell University Press, Ithaca, NY, 274.

Crane, E. 1992. The world's beekeeping-past and present. In: *The hive and the honey bee*, 1–22.

Crane, E. 1994. Beekeeping in the world of ancient Rome. *Bee World*, *75*(3), 118–134.

Crane, E. 1999. *The world history of beekeeping and honey hunting*. Routledge.

Crane, E. 2004. A short history of knowledge about honey bees (*Apis*) up to 1800. *Bee World*, *85*(1), 6–11.

Crane, E. and IBRA 2001. Traditional beekeeping in Asia: Some known and unknown aspects of its history. In *Proceedings of the 7th International Conference on Tropical Bees: Management and Diversity and the 5th Asia Apicultural Association Conference*, 19–25.

Crane, E. and Walker, P. 1984. Pollination directory for world crops. *Int. Bee Res. Assoc.*, 184.

Dave, K.V., 1954. Beekeeping in ancient India. *India Bee J.*, *16*, 92–95, 149–161, 169–191, 196–205.

Dave, K.V., 1955. Beekeeping in ancient India. *India Bee J.*, *17*, 11–25, 49–63, 87–97, 115–125, 169–176, 189–200 and 202.

De Candolle, A.P. 1832. The life and writings of Francis Huber. *Edinburgh Philos. J. Publ. 14*, 283–296.

Deodikar, G.B. 1959. Some taxonomic problems in honeybees: I. The concept of supra-generic grouping. *Bee World*, *40*(5), 121–124.

Deodikar, G.B. 1978. Possibilities of origin and diversification of angiosperms prior to continental drift. In Recent Researches in Geology, 474–481. Hindustan Publishing Corp., Delhi, India.

Dietz, A. 1962. A short natural history of the honey bee family Apidae (Leach, 1817). *Australas. Beekeep.*, *63*, 187–188.

Dobbs, A., 1750. IV. A letter from Arthur Dobbs Esq; to Charles Stanhope Efq; FRS concerning bees, and their method of gathering wax and honey. *Philos. Trans. R. Soc. Lond.*, *46*(496), 536–549.

Doolan, R. 1995. Dancing bees, *Creation Ex Nihilo.*, *17*(4), 16–19.

Doolittle, G.M. 1915. Scientific queen-rearing as practically applied: being a method by which the best of queen-bees are reared in perfect accord with nature's ways. *Am. Bee J.* Hamilton, USA, 126.

Dyer, F.C. 1985. Nocturnal orientation by the Asian honey bee, *Apis dorsata*. *Anim. Behav.*, *33*(3), 769–774.

Ebert, A.W. 2009. Hive society: the popularization of science and beekeeping in the British Isles, 1609–1913. Graduate Theses and Dissertations. 10587https://lib.dr.iastate.edu/etd/10587

Engel, M.S. 1999. The taxonomy of recent and fossil honey bees (*Hymenoptera: Apidae; Apis*). *Journal of Hymenoptera Research*, *8*(2), 165–196. http://www.pensoft.net/journals/jhr/

Everett Mendelsohn; Garland E. Allen. 2002. *Science, history, and social activism.* Springer. ISBN 978-1-4020-0495-7.

Fasasi, K.A. 2016. Research Article Comparative Seasonal Yield of Colonies of *Apis mellifera* adansonii (*Hymenoptera: Apidea*) in Response to some Environmental Variables. *Journal of Entomology*, *13*(1), 11–18.

Free, J.B. 1987. *Pheromones of social bees.* Cornell Univ. Press, Ithaca, NY.

Free, J.B. 1993. *Insect pollination of crops.* Second Addition Academic press, Landon.

Gartziou-Tatti, A. 2012. Haralampos V. Harissis, Anastasios V. Harissis, Apiculture in the Prehistoric Aegean. Minoan and Mycenaean Symbols Revisited. *Kernos. Revue internationale et pluridisciplinaire de religion grecque antique*, *1*(25), 334–335.

Gauld I. and Bolten B. 1988. *The hymenoptera*. British Museum (N.H.) and Oxford University Press, 20 (4), 332.

Gerstäcker, A. 1863. XXXI.—On the geographical distribution and varieties of the Honey-Bee, with remarks upon the Exotic Honey-Bees of the Old World. *Mag. Nat. Hist.*, *11*(64), 270–283.

Gojmerac, W.L. 1980. *Bees, beekeeping, honey and pollination*. AVI Publishing Co, Westport, CT, 192.

Graham, J.M. (ed) 1992. *The hive and the honey bee Hamilton, III.* Dadant and Sons, Hamilton, Illinois, 1324.

Graham, J.M. 1992. The hive and the honey bee (No. 638.1 H5/1992).

Halcroft, M.T., Spooner-Hart, R., Haigh, A.M., Heard, T.A. and Dollin, A. 2013. The Gojmerac, W.L. 1980. In: *Bees, beekeeping, honey and pollination*. AVI Publishing Co., Westport, CT, 192.

Harissis (Χαρίσης), Haralampos (Χαράλαμπος); Mavrofridis, Georgios 2012. A 17th Century Testimony On The Use Of Ceramic Top-bar Hives. 2012 | Haralampos (Χαράλαμπος) Harissis (Χαρίσης) and Georgios Mavrofridis. *Bee World*, *89*(3), 56–58.

Harissis, H.V. and Mavrofridis, G. 2012. A 17th Century Testimony On The Use Of Ceramic Top-bar Hives. *Bee World*, *89*(3), 56–58.

Hatjina, F., Mavrofridis, G. and Jones, R. (eds), 2018. *Beekeeping in the Mediterranean from antiquity to the present*. Division of Apiculture.

Haydak, M.H. 1935. Brood rearing by honeybees confined to pure carbohydrates diet. *J. Econ. Entomol.*, *2*, 657–660.

Haydak, M.H. 1943. Larval food and development of caste in the honeybees. *J. Econ. Entomol.*, *36*, 778–792.

Haydak, M.H. 1970. Honey bee nutrition. *Annu. Rev. Entomol.*, *15*(1), 143–156.

Hays G.W.J. 1984. Supplementary feeding of honeybees. *Am. Bee J.*, *124*, 35–37, 108–109.

Hoover S.E.R., Keeling C.I., Winston M.L. and Slessor K.N. 2003. The effect of queen pheromones on worker honey bee ovary development. *Naturwissenschaften*, *90*, 477–480.

Hossain, K.S., Hossain, M.G., Moni, A., Rahman, M.M., Rahman, U.H., Alam, M., … Uddin, M.J. 2020. Prospects of honey in fighting against COVID-19: pharmacological insights and therapeutic promises. *Heliyon*, *6*(12), e05798.

Huber F. 1814. New Observations Upon Bees, Hamilton, Illinois, Am. Bee J., translated by Dadant C.P., 1926.

Huber, F. 1792. New observations on bees, I and II. *Transl. CP Dadant Dadant Sons Hamilt.*, *111*, 230.

Jayasvati, S. and S. Wongsiri 1993. Scanning electron microscopy analysis of honey bee stings of six species (*Apis florea, Apis dorseta, Apis cerana, Apis koschevnikovi, Apis florea,* and *Apis andreniformis*). *Honeybee Sci.*, *14*, 105–109.

Johansson, T.S. 1955. Royal jelly. *Bee World*, *36*(2), 21–32.

Johansson, T.S.K. and Johansson, M.P. 1967. Lorenzo L. Langstroth and the bee space. *Bee World*, *48*(4), 133–143.

Kasangaki, P., Chemurot, M., Sharma, D. and Gupta, R.K. 2014. Beehives in the world. In *Beekeeping for poverty alleviation and livelihood security*. Springer, Dordrecht, 125–170.

Kilani, M. 1999. Biology of the honeybee. In: Colin M.E. (ed.), Ball B.V. (ed.), Kilani M. (ed.). Bee disease diagnosis. Zaragoza: CIHEAM, 9–24 (Options Méditerranéennes: Série B. Etudes et Recherches; n. 25)

Koeniger, G., Mardan, M., Otis, G. and Wongsiri, S. 1991. Comparative anatomy of male genetalia organs in the genus *Apis Apidologie.*, *22*, 539–552.

Koeniger, G. 1986a. Reproduction and mating behavior. *Bee Genet. Breed.*, Academic Press, San Diego, 255–280.

Koeniger, N., Weiss, J. and Maschwitz, U. 1979. Alarm pheromones of the sting in the genus Apis. *J. Insect Physiol.*, *25*(6), 467–476.

Koutchoumoff, Lisbeth, November 16, 2018. L'étonnanteHistoire du Genevois François Huber, apiculteuraveugle et visionnaire. *Le Temps.*

Kremen, C., Williams, N.M. and Thorp, R.W., 2002. Crop pollination from native bees at risk from agricultural intensification. *Proc. Natl. Acad. Sci. U.S.A.*, *99*, 16812–16816.

Kritsky, G. 2015. *The tears of Re: Beekeeping in ancient Egypt.* Oxford University Press.

Kritsky, G. 2017. Beekeeping from antiquity through the Middle Ages. *Annu. Rev. Entomol.*, *62*, 249–264.

Laidlaw, H.H. 1992. Production of queens and package bees. In: *The hive and the honey bee.* 989–1042.

Langstroth, L.L. 1852. *Beehive.* United States Patent Office. Letters Patent, (9,300).

Langstroth, L.L. 1853. *Langstroth on the hive and honey-bee, a bee keeper's manual.* Hopkins, Bridgeman and Co., Northampton, MA.

Langstroth, L.L. 1857. *A practical treatise on the hive and honey-bee.* CM Saxton & Company.

Lin, H. and Winston, M.L. 1998. The role of nutrition and temperature in the ovarian development of the worker honey bee (*Apis mellifera*). *Can. Ent.*, *130*, 883–891.

Lindauer, M. 1957. Communication among the honeybees and stingless bees of India. *Bee World*, *38*(1), 3–14.

Loper, G.M. and Berdel, R.L. 1980. The effects of nine pollen diets on broodrearing of honeybees. *Apidologie*, *11*(4), 351–359.

Maa, T.C. 1953. An inquiry into the systematics of the tribus Apidini or honeybees (Hym.). *Treubia*, *21*(3), 525–640.

Maderspacher, F. 2007. All the queen's men. *Curr. Bio.*, *17*(6), 191–195.

Malaka, S.L.O. and Fasasi, K.A. 2005. A review of beekeeping in Lagos and its environs. *Nigerian J. Entomol.*, *22*, 108–117.

Mason, I.L. and Mason, I.L. (eds), 1984. *Evolution of domesticated animals*, Longman, United Kingdom.

Matthews, Robert W. 2017. "The Wasps. Howard E. Evans and Mary Jane West Eberhard, with drawings by Sarah Landry. Ann Arbor: The University of Michigan Press, 1970. vi, 265. Paper, $3.45.," The Great Lakes Entomologist, *4*(1). https://scholar.valpo.edu/tgle/vol4/iss1/3

Mavrofridis, G. 2016. Traditional beekeeping in Crete (17th–20th century). In: Proceedings 12th International Congress of Cretan Studies, Hearklion, https://www.academia.edu/39058784/Traditional_beekeeping_in_Crete_17th_20th_century_

Mazar, A. and Panitz-Cohen, N. 2010. It is a land of honey.

McIndoo, N.E. 2017. *The auditory sense of the honey-bee.* Read Books Ltd.

Michener, C.D. 1978. The parasitic groups of Halictidae (Hymenoptera, Apoidea). *Univ. Kans. Sci. Bull.*, *51*(10), 291.

Michener, C.D. 2000. *The bees of the world.* Johns Hopkins University Press, Baltimore, MD.

Michener, C.D. and Michener, C.D. 1974. *The social behavior of the bees: A comparative study*. Harvard University Press.

Mishra, R.C. 1995. *Honeybees and their Management in India*. ICAR Publ., New Delhi, 168.

Mishra, R.C. and Sharma, S.K. 1997. Technology for management of *Apis mellifera* in India. *Perspective in Indian apiculture. Agro Botanica, New Delhi*, 131–149.

Morgenthaler, O. 1931. For the centenary of Huber, the blind investigation of the bee. *Bee World, 12*(12), 135–138.

Morison, G.D. 1927. The muscles of the adult honeybees *(Apis mellifera* L.). *Quart. J. Mocr. Sci., 71*, 395–463.

Moritz, B. and Crailsheim, K. 1987. Physiology of protein digestion in the midgut of the honeybee (*Apis mellifera* L.). *J. Insect Physiol., 33*(12), 923–931.

Morse, R.A. 1979. *Rearing queen honey bees*. Wicwas Press, Ithaca, NY.

Morse, R.A. and Flottum, K. (eds), 1997. Honey bee pests, predators and disease. 3rd ed. A I Root Co, Medina, Ohio.

Morse, R.A., Shearer, D.A., Boch, R. and Benton, A.W. 1967. Observations on alarm substances in the genus Apis. *J. Apic. Res., 6*(2), 113–118.

Narayanan, E.S., Sharma, P.L., and Phadke, K.G. 1961. Studies on requirements of various crops for insect pollinators, *Indian Bees J., 22*, 7–11 and 1960, 23, 23-30.

Odinsson, E. 2010. *Northern Lore: A field guide to the northern mind-body-spirit*. Eoghan Odinsson.

Palatty, A.S. and Shivanna, K.R. 2007. Pollination biology of large cardamom *(Amomum subulatum). Curr. Sci., 93*(4), 548–552.

Paliwal, G.N. and Tembhare, D.B. 1998. Surface ultra-structure of the sting in the Rock Honeybee *Apis dorsata (F.)* (Hymenoptera: Apidea), *Entomone, 23* (3), 203–209.

Park, C. 1925. Skep Beekeeping: Research, Resurgence and Resilience. http://www.dave-cushman.net/bee/Chris%20Park%20Article%202.pdf

Pattinson, D. 2012. Pre-modern beekeeping in China: A short history. *Agric. Hist., 86*(4), 235–255.

Paul, I.M., Beiler, J., McMonagle, A., Shaffer, M.L., Duda, L. and Berlin, C.M. 2007. Effect of honey, dextromethorphan, and no treatment on nocturnal cough and sleep quality for coughing children and their parents. *Arch. Pediatr. Adolesc. Med., 161*(12), 1140–1146.

Pernal, S.F. and Currie, R.W. 2000. Pollen quality of fresh and 1-year-old single pollen diets for worker honey bees (*Apis mellifera* L). *Apidologie, 31*, 387–409.

Preston, C. 2006. *Bee*. Reaktion Books, London.

Ranz, R.R. 2020. Introductory chapter: Actuality and trends of beekeeping worldwide. In *Beekeeping-New Challenges*. IntechOpen.

Raven, P.H. and Axelrod, D.I. 1974. Angiosperm biogeography and past continental movements. *Ann. Mo. Bot. Gard., 61*(3), 539–673.

Rawat, B.S. 1982. Bee Farming in India, Rawat Apiaries, Ranikhet. *Keywords: beekeeping/ cerana/dorsata/India*.

Ren, R. 2020. *Comb the honey: bee interface design* (Doctoral dissertation, Massachusetts Institute of Technology).

Requier, F., Rome, Q., Villemant, C. and Henry, M. 2020. A biodiversity-friendly method to mitigate the invasive Asian hornet's impact on European honey bees. *J. Pest Sci., 93*(1), 1–9.

Ribbands, C.R. 1953. *The behaviour and social life of honeybees* (No. QL568. A6 R48).

Richards, O.W. 1953. *The social insects* (No. QL569. R52 1970.). London: Macdonald.

Riedel Jr, S.M. 1967. Development of American beehive. *Beekeeping U.S.A.*, Agriculture Handbook No. (335), 8.

Robinson, W.S.R., Nowogrodzki, R. and Morse, R.A. 1989. The value of honey bee as pollinator of US crops. *Am. Bee J.*, *129*, 411–423 and 477–487.

Roffet-Salque, M., Regert, M., Evershed, R.P., Outram, A.K., Cramp, L.J., Decavallas, O., ... Zoughlami, J. 2015. Widespread exploitation of the honeybee by early Neolithic farmers. *Nature*, *527*(7577), 226–230.

Root, A.L. 1993. *The ABC and XYZ of bee culture*. Edward Arnold publication Ltd., London.

Roubik, D.W. 1995. Pollination of cultivated plants in the tropics. In: *FAO Agric. Serv. Bull. No. 118*, FAO, Rome.

Roy, P. 1999. Bee-diversity across a tropical tract-honey bees and people in India—An overview. *Beekeeping Dev.*, *52*, 3–4.

Ruttner, F. 1975. *The instrumental insemination of the queen bee* (No. SF 531.5. I57 1976).

Ruttner, F. 1986. Geographical variability and classification. In: *Bee genetics and breeding* (Rinderer, T.E., ed.). Academic Press, 23–56.

Ruttner, F. 1988. Breeding techniques and selection for breeding of the honeybee. British Isles Bee Breeders' Association.

Sakagami, S.F., Matsumura, T. and Ito, K. 1980. *Apis laboriosa* in Himalaya, the little-known world's largest honeybee (hymenoptera, Apidae). *Insect Matsumumara*, *19*, 47–77.

Saranraj P., Sivasakthi, S. and Feliciano, G.D. 2016. Pharmacology of honey—A review. *Adv. Biol. Res.*, *10*(4), 271–289.

Savoor, R.R. 1996. Managed pollination – An eco-friendly green revolution by-passes India. *Natl. Bank Newsl.*, *5*(6), 1–3.

Schäfer, M.O., Dietemann, V., Pirk, C.W.W., Neumann, P., Crewe, R.M., Hepburn, H.R., Tautz, J. and Crailsheim, K. 2006. Individual versus social pathway to honeybee worker reproduction (*Apis mellifera*): pollen or jelly as protein source for oogenesis? *J. Comp. Physiol. A.*, *192*(7), 761–768.

Schemske, D.W. 1981. Floral convergences and pollinators sharing on two bees-pollinated herbs. *Ecology*, *62*, 946–954.

Schmidt, L.S., Schmidt, J.O., Rao, H., Wang, W. and Xu, L. 1995. Feeding preference and survival of young worker honey bees (Hymenoptera: Apidae) fed rape, sesame, and sunflower pollen. *J. Econ. Entomol.*, *88*(6), 1591–1595.

Seeley, T.D. and Morse, R.A. 1976. The nest of the honey bee (*Apis mellifera* L.). *Insectes Sociaux*, *23*(4), 495–512.

Seeley, T.D. and Visscher, P.K. 1985. Survival of honeybees in cold climates: The critical timing of colony growth and reproduction. *Ecol. Entomol.*, *10*(1), 81–88.

Shende, S.G. and Phadke, R.P. 1993. History of beekeeping in India present status and scope. In *Proceedings of 1st National Conference of Beekeeping*, 9–21.

Simpson, J. 1960. Male genetalia of Apis species. *Nature*, *185*, 56.

Singh, A.K. 2014. Traditional beekeeping shows great promises for endangered indigenous bee *Apis cerana*.

Singh, S. 1975. Enemies and diseases of honey bee. *Beekeeping in India*, ICAR, New Delhi, 166–168.

Sivaram, V. 2012. Status, prospects and strategies for development of organic beekeeping in the South Asian Countries. In: *Division of apiculture and biodiversity*. Department of Botany, Bangalore University.

Smith, F.G. 1960. Beekeeping in the tropics. *Beekeeping in the tropics*.

Smith, F.G. 1961. The races of honeybees in Africa. *Bee World*, *42*(10), 255–260.

Snodgrass, R.E. and Erickson, E.H. 1992. The anatomy of the honeybee, In: *The hive and the honeybee* (Graham, J.M., ed). Dadant and Sons, Hamilton, IL, 103–169.

Snodgrass, R.E., 1925. *The anatomy and physiology of the honeybee*. McGraw-Hill Book Co., New York, 327.

Somerville, D.C. 1997. Value of pollens collected from agriculture crops. *Proceedings Crops Pollination Association Conference*. Tatura, Vic, 14–15 August.

Southwick, L.J.R. and Southwick, E.E. 1992. A comment on 'value of honey bees' as a pollinator of U.S. crops. *Am. Bee J.*, *129*, 805–807.

Stitz, J. and Szebe, P. 1933. The evolution of the races of bees and their geographical distribution. *Bee World*, *14*, 128–130.

Suryanarayana, M.C., Subba Rao, K. 1997. *A. cerana* for Indian Apiculture and its management technology: Perspective *in Indian apiculture: Agro Botanica*, 66–118.

Taber, S. and Owens, C.D. 1970. Colony founding initial nest design of honey bee (*Apis mellifera* L.).*Anim. Behav.*, *18*, 625–632.

Taber, S. and Wendel, J. 1958. Concerning the number of times queen mate. *J. Econ. Entomol.*, *51*, 786–789.

Tantawy, M. 2020. Efficacy of Natural Honey Treatment in Patients with Novel Coronavirus. Clinical Trials. gov.

Tej, M.K., Aruna, R., Mishra, G. and Srinivasan, M.R. 2017. Beekeeping in India. *Int. J. Ind. Entomol.*, 2017, 35–66.

Thomas, D., Pal, N. and Rao, K.S. 2002. Bee management and productivity of Indian honeybees. *Apiacta*, *3*, 1–15.

Traynor, K. 2008. Ancient Cave Painting Man of Bicorp. (Web article). MD Bee. Retrieved.

Verma, L.R. 1991. Beekeeping in integrated mountain development. *Beekeeping in integrated mountain development*. Aspect Publications Ltd, United Kingdom, 384.

von Frisch, K. 1971. *Bees, their vision, chemical senses, and language*. Cornell University Press. Ithaca, NY.

Wallace, V.A. 2018. *Buddhist medicine in India*. In: Oxford research encyclopedia of religion.

Weaver, N. and Weaver, E.C. 1981. Beekeeping with the stingless bee *Meupona beecheii*, by the Yucatecan Maya. *Bee World*, *62*(1), 7–19.

Wedmore, B. 1979. *A manual of beekeeping*. Bee Books New and Old, Burrowbridge, UK.

Wheeler D. 1996. The role of nourishment in oogenesis. *Ann. Rev. Entomol.*, *41*, 407–431.

White, J.W. 1967. Honey, its composition and properties. *Beekeeping in the United States. Handb. US Dep. Agric.*, *335*(147), 56–64.

Wildman, T. 1768. A Treatise on the Management of Bees. author, and sold.

Wilson, W.T. 1971. Resistance to American foulbrood in honey bees. XI: Fate of Bacillus larvae spores ingested by adults. *J. Invertebr. Pathol.*, *17*(2), 247–255.

Winston, M.L. and Michener, C.D. 1977. Dual origin of highly social behavior among bees. *Proc. Natl. Acad. Sci.*, *74*(3), 1135–1137.

Wintson, M.L. 1987. *The biology of honey bees*. Harvard University Press, London.

Woyke, J. 1976. Brood-rearing efficiency and absconding in Indian honeybees. *J. Apic. Res.*, *15*(3–4), 133–143.

Zeuner, F.E. and Manning, F.J. 1976. A monograph on fossil bees (*Hymenoptera: Apoidea*). British Museum, London, United Kingdom.

Zheng, H., Cao, L., Huang, S., Neumann, P. and Hu, F. 2018. Current status of the beekeeping industry in China. In *Asian beekeeping in the 21st century*. Springer, Singapore, 129–158.

2

Bee Flora and Biology of Honey Production

Jasvir Singh Dalio
G.G.S.S.S Budhlada, Budhlada, India

CONTENTS

Introduction

The symbiosis of anthophilous insects like honeybees and entomophilous flowering plants has evolved in millions of years. These angiosperms have developed highly specialised flowers with rewards (nectar and pollen) to attract bees, which increases the reproductive capabilities of plants through pollination. This symbiotic relationship between these distantly related taxa has brought about a high degree of complexity and diversity. Honeybees show structural, physiological and behavioural adaptations to get maximum nutritional benefits offered by blossoms, which are the mainstay of their lives. They are micromanipulators by the help of which we can manipulate floral resources which are otherwise unobtainable. As they pollinate a wide range of species of angiosperms while collecting pollen and nectar, they are considered responsible for one out of every three bites of food humans eat. Flowers are patchily distributed in time and space, and honeybees locate them wonderfully and visit them by adopting systematic foraging strategies. Honeybees also collect propolis or bees glue from various plants and use it to protect the colony from pathogens and intruders and maintain physical conditions inside the hive. The co-evolution of floral characteristics of angiosperms with honeybees and other pollinators is considered a primary reason for morphological divergence and speciation in flowering plants. Thus, *Apis* species and flowering plants evolved a well-adjusted system of interdependence, which is amazing event of organic evolution. The insects that specialise in visiting specific flowers have developed an array of foraging strategies to improve their efficiency (Goulson, 1999). Nectar plays an essential role in attracting insect pollinators (Leiss *et al.*, 2004; Kudo and Harder, 2005), and nectar production is believed to be a part of evolutionary development (Hooper, 2009; Wesselingh and Arnold, 2000), but it is also affected by environmental factors (Shuel, 1992; Macukanovic-Jocic and Djurdjevie, 2005; Phillips *et al.*, 2010). *Apis* species show a unique foraging behaviour that may have evolved more than 20 million years ago. The majority of floral resources in India are still under or unutilised due to a lack of knowledge and related research work. Beekeeping depends not only on the better strain of *Apis mellifera* but also on the status of bee pasture (nectar or pollen plants) existing in a given area, as apiculture is agro-horticultural and forest-based industry. The plant species, which support the growth and development of the honeybee colonies in terms of providing nectar or pollen, are called bee flora. With the background mentioned earlier, the main aim of this chapter is to provide information regarding bee flora, floral rewards, foraging behaviour of honeybees on different flowering plants, evaluation of bee plants, the status of other flower species as bee flora, floral calendar, the biology of honey production and future strategies for management of bee flora for promoting beekeeping and sustaining of wild *Apis* species (*Apis cerana*, *Apis florea* and *Apis dorsata*) by protecting them from nutritional stress.

Bee Flora

Honeybees do not visit the bloom of every plant species as quantity and quality of forage may be unsatisfactory or even negligible in some flowers, or bee foragers are unable to collect rewards due to inappropriate floral structure. Bees have remarkable aptness to differentiate among the flowers with varying amounts of tips without actually visiting them (Heinrich, 1979; Marden, 1984; Wetherwax, 1986; Kato, 1988; Goulson *et al.*, 2001). They use morphological cues like colour, size, ultraviolet reflectance and floral symmetry to discriminate among them (Inouye, 1983; Waser, 1983; Kearns and Inouye, 1993; Conner and Rush, 1996; Moller and Sorci, 1998; Elle and Carney, 2003). To fulfil their nutritional requirements, they collect pollen and nectar from different species of angiosperms, including fruit, vegetable, oilseed, ornamental, medicinal, spice, weed, wild or avenue plants, along with many crops.

These flowering plants which provide forage to honeybees are termed bee flora. These plants may be classified as colony building, honey flow and subsistence bee flora depending upon the floral rewards and blooming season.

Collection of forage is a continuous process throughout the year, so the sequence of flowering in the vicinity of the apiary is essential, enabling bees to shift from one type of bloom to another to exploit various floral sources without any gap. An abundance of flowers (more area under flowering crop) is another factor to attract more bee foragers than small patches of flowers. Good nutrition in terms of large bee florane apiary, around the year, has been identified as a critical factor to maintain health and flourishment of the bee colonies for commercial apiculture and conservation of wild honeybees. Deforestation, unsustainable agriculture, lack of crop diversification, shrinking of natural areas and unplanned development are causing nutritional stress; thus, the honeybee carrying capacity of various ecosystems decreases, particularly in developing countries. Bee flora varies from area to area and during different periods of the year. Every region has its periods of floral abundance and floral scarcity. It is necessary to fill the gaps of floral dearth by identifying and growing suitable bee plants.

The localities with many plant species having blossoms rich with pollen and nectar for a long time are suitable for beekeeping. Still, every ecosystem has periods of the floral paucity of long or short duration, which is a critical problem in apiculture. Sometimes flowers are rich with rewards which are not accessible for bees due to obstructions of floral morphology. For example, flowers of many cultivars of *Aloe vera* have a large quantity of high-quality nectar. Still, bee foragers cannot get it due to the long and narrow corolla tube obstructed by androecium. The status of bee flora in any area determines the bee colony carrying capacity and potential of honey production. Knowledge of plants of bee interest is essential for successful beekeeping as the presence of such flowers in abundance directly affects colony performance and honey yield. More information regarding nectariferous and polleniferous plant sources of the area during various periods of the year provides guidelines to standardise the management practices of the apiaries.

For selecting a good apiary site, bee forage is the first thing to know and crop rotation pattern. Not only primary source but also the presence of minor and subsistence flora ensure the sustaining of bee colonies. Sometimes plants of bee interest are present even in small pockets of the given area, as some flowering plants are specific to particular habitats. Other aspects of plant which should be explored are taxonomy, geographic distribution, the status of floral rewards (quantity and quality of nectar and pollen), blooming season, length of the flowering period, degree of attractiveness of bees to its flowers, flower density, foraging behaviour of *Apis* species on its bloom and feasibility in taking rewards (suitable floral morphology). Such information helps to adjust colony manipulation tasks following the blooming time of local vegetation.

Production of honey and other hive products depends on the availability of bee pasture, an essential field of research for biologists and apiculturists. The blooming of plants related to various families occurs during different months of the year, so the availability of floral rewards and their collection by honeybees fluctuates with time. When major nectar plant species are blooming abundantly near apiaries, they cause honey flow, as nectar collection causes honey production. In North India, two to three honey flows have been recorded in a year, but monoculture of the crops that are not of bee interest causes long dearth periods in some areas. Crop diversification of the bee interest may help a lot in such regions. Honey flow season varies from one agro-climatic zone to another, so beekeepers should do a survey and study floral calendars of different areas to do migratory beekeeping to increase honey yield, keep the honeybee colonies robust and make this small cottage industry more profitable as such management ensures continuous supply of pollen and nectar.

Nectar

Blooming bee flora is the lifeline of honeybees, which provides them food in nectar and pollen. Nectar is a carbohydrate-rich secretion, which acts as a fuel for performing all the life activities of the bees and raw material for honey production. Floral or extrafloral nectaries of plants secrete nectar. It is mainly a solution of different types of sugars like glucose, fructose, sucrose and maltose, minute quantities of amino acids, minerals and aroma oils. Honeybees show more attraction to the flowers yielding nectar with higher sugar concentration, and the degree of attractiveness further increases when glucose, fructose and sucrose are present in equal amounts. The ratio of these sugars in nectar remains constant for particulars plants species. Therefore, the composition of different sugars and their concentration in nectar are valuable factors for evaluating any flower species as bee flora.

Variations in Quality and Quantity of Nectar

Nectar sugar concentration in most bee flower species varies from 20 to 50% (Mishra, 1995). However, bees may visit the bloom with nectar having 6–79% sugar content in different situations (Deodikar et al., 1957). However, when this concentration is less than 20%, net energy gain is almost negligible, but bees may forage relatively poor floral resources under nutritional stress during dearth periods.

The quality and quantity of nectar vary not only with plant species but also in different varieties/cultivars of the same species. Amount and sugar concentration level of the nectar of a specific cultivar of a crop may also change with variation in agricultural practices, level of growth hormones, age of flowers and environmental factors. Atmospheric temperature, relative humidity and soil factors like aeration, water contents and nutrient level are the external factors which influence the quantity and quality of nectar. Day-hour-related floral physiology of blooming plants also affects the floral rewards, and bees generally visit them when floral tips are at the peak.

Morphology of flower surface area of nectaries as well as qualitative and quantitative properties of nectar is controlled by the genetic makeup of the plant species or different cultivars of the same species. However, variation in the nectar quantity has also been observed (Percival, 1946) with a change in location and perimeter of the flowering branch of the same tree. The difference in floral rewards has also been observed in male and female flowers of the same plant. Nectar secretion is further stimulated by the frequent visitation of honeybees (Percival, 1965; Mishra *et al.*, 1985; Mishra and Sharma, 1989).

Pollen

Pollen is a protein-rich material, which also contains vitamins, fat and other nutrients. It gets produced within the anthers of flowers. Honeybees require pollen for brood rearing. The amount of brood rearing is primarily determined by the quantity and quality of pollen available, as one cell of pollen is necessary to rear one larva. Pollen shortage is calamitous for bee colonies, and bee strength dwindles significantly as old and depreciated bees die and new brood is not reared. It is necessary to fulfil the pollen requirements of the colonies to build up and make them robust to collect more honey. The pollen requirements of a well-flourishing colony are 15–30 kg/year, and pollen load per bee varies from 14 to 20 mg. The colour of stored pollen in combs helps to identify the floral source (Figures 2.1 and 2.2).

Pollen Collection

Foraging activity in the field depends on the colony's requirements and the availability of floral rewards (physiology of flowering plants). During the breeding season of honeybees, brood rearing increases, raising pollen requirement and enhancing pollen-collecting activity. Releasing pollen grains from anthers (dehiscence) is a circadian rhythm that takes place at fixed hours of the day; thus, characteristic of a plant species and honeybees regulate their foraging activity accordingly. Many researchers (Levin and Bohart, 1955; Doull, 1966; Free, 1993) have shown that certain chemicals

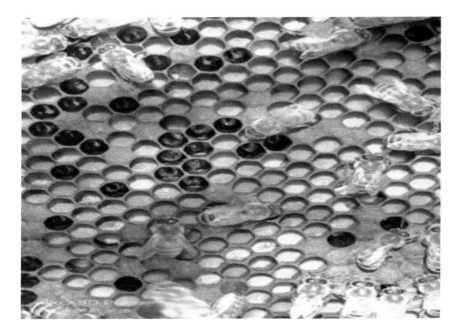

FIGURE 2.1   Pollen stores of *Apis mellifera*, collected from different flower spp.

FIGURE 2.2   Pollen of different colours in comb.

in pollen grains are responsible for making pollen of particular plant species attractive to bees and spraying of these chemicals on the desired crop may help to attract more bee foragers to collect more pollen and enhance pollination. A more significant number of variations are seen in pollen collection during different day hours and peak foraging in other species of plants. These differences are due to dissimilarity in time-related floral physiology of different plant species and bees to various stimuli.

Evaluating the Bee Flora

Evaluating the status of various plants in the locality, as bee flora and their potential to provide nectar and pollen, is an essential component for the development of beekeeping.

Many strategies to study bee flora in any ecosystem include:

- Grid matrix.
- Visual observations of flowering.
- Foraging behaviour of honeybee on the bloom of different species at different day hours.
- Microscopic analysis of pollen loads carried by foragers or pollen collected by using pollen traps hive entrances.

Melissopalynological study of honey samples in different periods of the year and their comparison with reference slides can also confirm floral sources. A regular record of bee colonies in the apiary is also a vital method to know the status of the bee flora during different months of the year. Blooming season and duration of flowering (long or short) also tell the value of plant species. Other aspects of the study to know the importance of flora include total nectar secretion per flower, nectar sugar concentration and total sugar content per flower, which help estimate total nectar production and nectar available or harvested by bees and give valuable data regarding honey production. The status of bee flora near bee yard can also be assessed by observing comb area under pollen stores, unripe and ripened honey, various stages of brood and bee strength of colonies at different times of the year.

Agricultural Crops as Bee Flora

Cotton (*Gossypium* spp.; Malvaceae) flowering is attractive to honeybees and takes place for a long time of 4 months. They visit the flowers to collect pollen as well as nectar from floral and extrafloral nectaries. Some other species of cotton like Cambodia cotton provide much nectar to honeybees which produce reddish-coloured honey from it in large amounts (Singh, 1962). Cotton honey is dense, having a delightful taste and light aroma. The bees do not visit cotton bloom if another attractive flora is present in the vicinity (Mishra, 1995).

Crops of *Brassica* spp. (Cruciferae) are grown extensively in India. However, the blooming of various brassicas species starts from October and continue up to mid-March, but the honey flow is from December to February. The bloom of these crops is attractive to all types of honeybees (Figures 2.3–2.6) due to characteristic scent,

FIGURE 2.3 *Apis mellifera* forager on *Brassica campestris* flowers.

FIGURE 2.4   *Apis florea* nectar gatherer on *Brassica campestris* bloom.

colour (yellow), excellent rewards and other morphological and physiological characteristics of flowers. The flowering time of the species like *Brassica napus* coincides
with the breeding time of honeybees, particularly *A. mellifera*, and pollen is required
in large quantities to rare maximum brood. This crop provides pollen and nectar of
higher quality in large amounts. Full foraging activity of honeybees has been noted
during 1200–1400 hours (Dalio, 2003).

FIGURE 2.5   *Apis cerana* pollen collector on *Brassica campestris* flower.

FIGURE 2.6 *Apis dorsata* foraging on *Brassica campestris* flowering.

Foraging activity of *A. mellifera* on *Brassica juncea* and *Brassica campestris* bloom is from 1100 to 1600 hours (Jhajj *et al.*, 1996). Surplus honey is obtained during January and February. *Brassica* honey is yellowish-white with some pungent flavour and granulates quickly (Mishra, 1995).

Sunflower (*Helianthus annuus*; Compositae) is an important oil seed crop and an ornamental plant. Nectar sugar concentration is 33–60% in different varieties/cultivars. The crop is an excellent source of pollen but a medium source of nectar. Pollen foraging is during the morning and evening. Honey is yellow or golden with moderate sweetness, short aftertaste and granulation is rapid (Mishra, 1995).

Egyptian clover (*Trifolium alexandrinum* L.; Leguminosae) is an essential crop in Egypt and introduced in northern India, especially Punjab and Haryana, in the early nineteenth century as a significant fodder crop. After taking four to five cuttings, the harvest is allowed to bloom as a seed crop. The flowering provides both nectars and pollen but is a significant nectar source and results in exceptional honey flow in April and May. Some cultivars show flowering up to mid-June that is a period of floral scarcity.

Trifolium sp. as a significant source of surplus honey has been confirmed by many workers (Atwal *et al.*, 1970; Naim and Phadke, 1976). The author himself observed peak foraging activity at midday (at 1200 hours) and maximum pollen foraging during morning hours.

The foraging frequency of *A. mellifera* bees was less (10–11 flowers/m²/min), indicating rich floral rewards. Generally, 15–17 kg honey/colony of *A. mellifera* is obtained if sufficient crop is present within the flight range of the apiary. Egyptian clover honey is light-coloured, having low water content and granulates very slowly.

Sesamum (*Sesamum indicum* L.; Pedaliaceae) is an important crop that blooms in July and August in northern India and has flowers of white colour with slight purple shade having floral and extrafloral nectaries; it is the major source of nectar as well as

FIGURE 2.7   *Apis florea* nectar collector on sesamum flower.

pollen for *Apis* species (Figures 2.7 and 2.8) during time of floral scarcity. Maximum foraging activity of *A. mellifera* is during morning hours and then decreases gradually as the day advances.

Complete pollen collection takes place between 900 and 1000 hours. Bloom can provide surplus honey and pollen to bee colonies (Taha *et al.*, 2018). Honey is light-coloured and granulates slowly.

Some other fodder and grain crops (family; Graminae) are a good source of pollen. Maize (*Zea mays*), pearl millet (*Pennisetum glaucum*), sorghum (*Sorghum bicolour*) are polliniferous crops. These are grown as fodder or grain crops. These plants provide a large amount of pollen when floral scarcity is prevailing. *Zea mays* produce large tassels which make much pollen. It is considered a medium source as pollen is dry and falls within about 2 hours (Mishra, 1995). Bloom of *P. glaucum* attracts more bees to provide pollen. Pollen collection by honeybee (*A. mellifera, A. cerana, A. dorsata* and *A. florea*) foragers can be observed in the morning during July to September. They visited the crop in such a vast number that humming sound was noticed in the field by the author. *Sorghum* species are also good pollen plants. Apart from the previous ones, some other essential crops related to bees are shown in Table 2.1.

FIGURE 2.8 *Apis mellifera* nectar seeker entering sesamum flower.

Fruit Plants as Bee Flora

Banana (*Musa* spp. Musaceae) blooms throughout the year, and these flowers are attractive to honeybees; more visitation took place during morning and evening. Flowers provide both nectar and pollen but good nectar supplier to bees, having 25–30% sugar concentration. Flowers are available even during dearth periods.

TABLE 2.1

Some Other Agricultural Crops of Bee Interest

Sr. No.	Name	Family	Source	Months	References
1	*Ammi majus*	Apiaceae	N+P	March–April	Abrol (2009)
2	*Arachis hypogea* L.	Caesalpiniaceae	N+P	Mid-June–August	Chaudhary (2003)
3	*Cajanus cajan*	Leguminosae	N+P	April–May	Abrol (2009)
4	*Cicer arietinum* L.	Leguminosae	N+P	February–March	Chaubal and Kotmire (1980)
5	*Dolichos lablab* L.	Caesalpiniaceae	N+P	August–October	Chaudhary (2003)

(Continued)

TABLE 2.1 *(Continued)*

Some Other Agricultural Crops of Bee Interest

Sr. No.	Name	Family	Source	Months	References
6	*Eruca sativa* Mill.	Crucifereae	N+P	January–April	Gupta *et al.* (1986)
7	*Fagopyrum esculentum* Moench	Polygonaceae	N+P	July–September	Saraf (1973)
8	*Glycine max* L.	Leguminosae	N	July–August	Atwal (2000)
9	*Guizotia abyssinica*	Compositae	N+P	September–October	Chaubal and Kotmire (1980)
10	*Lens esculenta* Moench	Caesalpiniaceae	P	January–mid-March	Abrol (2009)
11	*Linum usitatissimum* L.	Linaceae	N+P	December–February	Abrol (2009)
12	*Medicago sativa* L.	Leguminosae	N+P	March–May July–September	Atwal (2000)
13	*Nicotiana tabacum* L.	Solanaceae	N+P	November–April	Atwal (2000)
14	*Papaver somniferum*	Papaveraceae	P	February–March	Abrol (2009)
15	*Phaseolus mungo* L.	Leguminosae	N	August–September	Abrol (2009)
16	*Pisum sativum* L.	Caesalpiniaceae	P	January–mid-March	Chaudhary (2003)
17	*Ricinus communis* L.	Euphorbiaceae	N+P	February–March	Chaudhary (2003)
18	*Trifolium pratense* L.	Leguminosae	N+P	July–August	Atwal (2000)
19	*Vigna aconitifolia*	Leguminosae	N	March–May	Sivaram (2001)
20	*Vigna mungo*	Leguminosae	N	August–September	Sivaram (2001)
21	*Vigna radiata* (L.) Wilezek	Caesalpiniaceae	N+P	Mid-September–mid-October	Chaudhary (2003)
22	*Vigna unguiculata*	Leguminosae	N	July–August	Sivaram (2001)

Note: N=nectar, P=pollen, N+P=nectar and pollen.

According to an estimate (EL-Kazafy and Taha, 2007), one feddan of banana plantation can produce 65.13 kg of nectar/day, generating 16.51 kg of honey. Guava (*Psidium guajava* L., Myrtaceae) is a small fruit tree with attractive flowers, attractive to honeybees and a good source of nectar and pollen. The *Apis* species visit the bloom throughout the day, but foraging activity is maximum during morning hours and reduces gradually as the day advances. The flowering takes place in May and June, which is a period of floral scarcity. There are many varieties/cultivars, which bloom throughout the year. All types of honeybees visited guava flowers. The honey produced from its nectar is thin, pleasant tasting and light amber in colour.

Mango (*Mangifera indica* L.; Anacardiaceae) flowers are yellowish-green in terminal panicles. Pollen grains have much adhering property; thus, entomophily is essential. Honeybees visit the bloom for nectar as well as pollen. Blooming takes place in spring but at peak in March in northern India. It is a minor forage source, and surplus honey has been recorded occasionally. Foragers also collect honeydew from over-ripened fruits.

Blackberry (*Syzygium cumini*; Myrtaceae) is a significant bee flora in India's location and yields surplus honey (Kohli, 1958/59; Naim and Phadke, 1976; Chaudhary, 1977; Chaubal and Kotmire, 1980). Flowering takes place in April and May, and honey flow is for 2–3 weeks. It is dirty-white-coloured, fragrant and small flowers also provide pollen.

Honey produced is amber-coloured, has the peculiar natural aroma of the tree and shows no granulation.

Citrus spp. (Rutaceae) flowers are whitish to yellow or pink or purple in different species, having pleasing fragrance and attractive honeybees. They get both nectars as well as pollen from them. Blooming takes place during February and March, but certain species/cultivars bloom throughout the year. Nectar is produced copiously (20 µL/flower) from nectaries or floral disc, which continue to secrete for at least 2 days after the flower's opening (Vansell *et al.*, 1942). However, nectar sugar concentration is medium. Honeybees (Figures 2.9–2.11) visit lemon flowering during morning and evening with more intensity; however, foraging continues throughout the day.

This forage serves as a colony-building flora, but surplus honey is available only in certain pockets where its horticulture is done commercially over a large area. The specific markers of citrus honey are methyl anthranilate, hesperidin and caffeine.

It is light-coloured, having a mild taste with characteristic citrus flavour and fresh aroma. Litchi (*Litchi chinensis* Sonner; Sapindaceae) blooms in early spring (February and March), and self-sterile flowers have a plentiful amount of nectar which attracts honeybees and other insects for pollination. The flowers also provide pollen along with nectar. *Apis* species (*A. mellifera, A. cerana, A. dorsata, A. florea*) show peak foraging activity during morning hours, decreasing gradually as the day advances (Singh *et al.*, 2006). Many workers (Naim and Phadke, 1976; Chaudhary, 1977) have confirmed litchi bloom as a rich source of nectar for honeybees. The sugar concentration of nectar is high. Sucking of juice from damaged fruits has also been observed. Honey formed is light golden with the characteristic aroma of litchi.

Ziziphus mauritiana (Rhamnaceae) is a medium-sized tree with white or greenish-white-coloured flowers which are protandrous, desciflorous and rich with nectar. This flowering is attractive to all the species of honeybees (Figures 2.12 and 2.13). An average number of foragers of all the three bee species (*A. mellifera, A. cerana* and *A. florea*) on bloom were 13.42 (±2.51) bees/m²/min. During its full blooming (September to October) season, foragers of *Apis* species were seen foraging throughout the day, but peak foraging was recorded from 900 to1300 hours in the case of *A. florea* and from 930 to1030 hours both for *A. mellifera* and *A. cerana* (Dalio, 2018d).

The apiaries having *Z. mauritiana* trees in their vicinity showed exceptional honey flow during its blooming. Others (Abrol, 2009) have also reported more nectar production by its flowering.

Phyllanthus emblica L. (Phyllanthaceae) blooming takes place during April and May. Inflorescences is raceme type, and small-sized flowers are unisexual; thus, entomophily becomes essential for more fruit set. The bloom is attractive to honeybees and provides pollen and nectar to *A. mellifera* and wild species of *Apis*

FIGURE 2.9   *Apis mellifera* on lemon flower.

(*A. cerana* and *A. florea*). Maximum foraging takes place during morning and evening hours. In addition to the previous ones, some other horticultural plants are enlisted in Table 2.2.

Ornamental Plants as Bee Flora

Experiments (Dalio, 2017c) have revealed that *Jasminum grandiflorum* (Oleaceae) is a good source of nectar for *A. mellifera*, and its foragers show fascinating and specialised foraging behaviour as the reward is hidden near the lower end of the long and narrow basal tube. The foragers collected nectar when the thalamus became naked after shedding corolla tubes, and nectar was available to them. It seems that sweet

FIGURE 2.10 *Apis florea* visiting lemon flower.

fragrance helped the bees to detect flowers in time and space. The bloom may become significant to fill severe dearth gaps of summer. Nectar collected from this bloom results (if plants are present in plenty) in honey production with a pleasing odour of jasmine.

Observations (Dalio, 2017d) show that the Rangoon creeper (*Combretum indicum*; Cobretaceae) is in full bloom from March to August and is very attractive for domesticated (*A. mellifera*) and wild (*A. cerana* and *A. florea*) honeybees (Figures 2.14 and 2.15) and provides both nectar and pollen. Maximum abundance (5–6 bees/m^2/min) was seen during morning hours. The outer epithelial surface of the corolla tube also produced nectar droplets which were visible with a magnifying glass. Foragers visited upon that surface along with the front part of the flower to collect nectar. The study also reveals more availability of nectar as compared to pollen.

Growing this vine in houses, parks, farmhouses, avenues, kitchen gardens and around the apiaries may help sustain bee fauna.

FIGURE 2.11   *Apis cerana* forager on lemon bloom.

Cestrum nocturnum (Solanaceae) blooms from July to December. An investigation (Dalio, 2017e) of the foraging behaviour of honeybees on its bloom reveals it as a good source of nectar and pollen.

The maximum abundance of honeybees on bloom and total visitation for pollen and nectar has been recorded from 930 to 1130 hours. Pollen collecting bees of *A. cerana* and *A. mellifera* (Figures 2.16 and 2.17) were prevalent. *Apis* (*A. mellifera*, *A. cerana* and *A. florea*) visited the bloom frequently *and* were also seen foraging on shed corolla tubes. It provides forage to bee fauna at times when floral deficiency prevails. However, the plant blooms at night, but flowers remain open during next morning.

Research showed (Dalio, 2017f) that *Campsis tagliabuana* (Bignoniaceae), an ornamental plant having showy flowers, provided higher quality as well as quantity of nectar and pollen to honeybees during the short summer season (lean period of the year). *A. Umbelliferae* (Figure 2.18) showed strong attraction to these flowers.

FIGURE 2.12 *Apis florea* nectar seeker on *Ziziphus mauritiana* bloom.

FIGURE 2.13 *Apis mellifera* nectar gatherer on *Ziziphus* sp. bloom

TABLE 2.2

Some Other Fruit Plants of Bee Interest

Sr. No.	Name	Family	Source	Months	References
1	*Anacardium occidentale* L.	Anacardiaceae	N+P	February–March	Atwal (2000)
2	*Carica papaya* L.	Caricaceae	P	July–December	Chaudhary (2003)
3	*Citrus aurantifolia* (Christm) Swingle	Rutaceae	N+P	December–January	Srawan and Sohi (1985)
4	*Citrus deliciosa* Ten.	Rutaceae	N+P	February–March	Atwal (2000)
5	*Citrus medica*	Rutaceae	N+P	March, April	Abrol (2009)
6	*Citrus reticulata* Blanco	Rutaceae	N+P	March	Sharma and Raj (1985)
7	*Citrus sinensis* (L.) Osb.	Rutaceae	N+P	March	Srawan and Sohi (1985)
8	*Eriobotrya japonica* Lindl	Rosaceae	N	February–March August	Singh (1962)
9	*Fragaria ananassa* Duchesne	Rosaceae	N+P	December–March	Chaudhary (2003)
10	*Grewia asiatica* L.	Tiliaceae	N	April–August	Singh (1962)
11	*Grewia oppositifolia* Roxb.	Tiliaceae	N	April–May	Singh (1962)
12	*Malus domestica* Borkh	Rosaceae	N+P	March–April	Singh (1962)
13	*Phoenix dactylifera* L.	Palmae	N+P	June–July	Chaudhary (2003)
14	*Prunus amygdalus*	Rosaceae	N+P	February–March	Sharma and Gupta (1993)
15	*Prunus armeniaca* L.	Rosaceae	N+P	February–March	Singh (1962)
16	*Prunus puddum* Roxb.	Rosaceae	N+P	November–December	Gupta *et al.* (1990)
17	*Punica granatum* L.	Punicaceae	P	March–May	Chaudhary (2003)
18	*Pyrus communis* L.	Rosaceae	N+P	February–March	Sharma and Gupta (1993)
19	*Pyrus pashia* Bichham	Rosaceae	N+P	February–March	Gupta (1993)
20	*Pyrus persica* L.	Rosaceae	N+P	February–March	Sharma and Gupta (1993)
21	*Psidium guajava* L.	Myrtaceae	N+P	April–May September–December	Atwal (2000)
22	*Rubus* spp.	Rosaceae	N+P	March, April	Singh (1962)
23	*Ziziphus oxyphylla* Edgew	Rhamnaceae	N+P	July–August	Singh (1962)

Note: N=nectar, P=pollen, N+P=nectar and pollen.

FIGURE 2.14 Italian honeybee visiting flowers of *Combretum indicum.*

FIGURE 2.15 Indian honeybee foraging on Rangoon creeper.

FIGURE 2.16   *Apis cerana* on *Cestrum nocturnum* bloom.

FIGURE 2.17   *Apis mellifera* pollen forager on *Cestrum nocturnum* flowers.

FIGURE 2.18 *Apis mellifera* on *Campsis tagliabuana* flowers.

The abundance of bees showed two peaks, during the morning (1000–1200 hours) and evening (1600–1700 hours). On an average, 20 ±3.4 bees/m²/min were present. Two bees/flower were primarily reported, one gathering pollen while the other collecting nectar deep in the bud. They foraged so vigorously and voraciously that humming sound was audible from a distance. Foraging took place even on fallen corolla tubes under vines (Figure 2.19).

A study on the foraging behaviour of *A. mellifera* and *A. florea* on *Jatropha integerrima* (Euphorbiaceae) flowers revealed that this ornamental plant provided good forage (pollen and nectar) to them. Peak foraging activity was recorded from 800 to 900 hours. However, visitation continued throughout the day. The percentage of pollen gatherers was significantly more as compared to that of nectar collectors. (Dalio, 2017a). Generally, most of the its cultivars bloom throughout the year.

Tecoma castanifolia (Bignoniaceae) flowers attracted all types of honeybees (*A. mellifera, A. cerana and A. florea*), but Italian and dwarf bees visited more frequently (Figures 2.20 and 2.21). An average number of bees/m²/min on its bloom was more in the case of *A. mellifera* (5.7±0.52) followed by *A. florea* (2.88±0.40) and *A. cerana* (1.68±0.25). Foraging hours/day were 12, 11 and 9 for *A. mellifera*, *A. cerana* and *A. florea*, respectively. The foraging speed *of A. cerana* was maximum, followed by *A.*

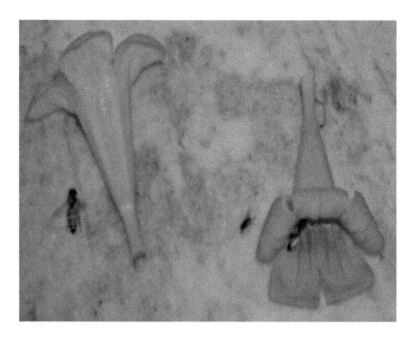

FIGURE 2.19   *Apis mellifera* collecting forage from shed corolla tubes.

FIGURE 2.20   Italian bees foraging on *Tecoma castanifolia* flower.

FIGURE 2.21 *Apis florea* forager on *Tecoma castanifolia* bloom.

mellifera and *A. florea*. The flowers are an excellent source of nectar during harassing time of summer.

A higher abundance of honeybees and more time spent/flower show that the experimental plant is good bee flora. Its many cultivars bloom almost throughout the year, especially during dearth periods (Dalio, 2021a).

Spermadictyon suaveolens Roxb (Rubiaceae) flowers have the potential to attract and feed honeybees. Visiting of *A. mellifera* (Figure 2.22) on its bloom continued throughout the day and was more during evening hours. All the foragers were nectar collectors; such observations clearly showed that bloom was a good source of nectar (Dalio, 2021b). Nectar seekers of *A. florea* (Figure 2.23) also visited the bloom frequently.

If grown in a large number at suitable places, the shrub may help as a subsistent nectar source for *A. mellifera* and *A. florea*. Furthermore, these shrubs are aesthetically pleasing and having medicinal properties.

The author observed the foraging behaviour of honeybees on *Dahlia* spp. (Figure 2.24), *Papaver rhoeas* (Figures 2.25 and 2.26), *Portulaca grandiflora*, *Glebionis coronaria* as well as *Antigonon leptopus* and revealed that open-centred

FIGURE 2.22 *Apis mellifera* on *Spermadictyon suaveolens flowers.*

(single) *Dahlia* spp. (Asteraceae) were a good source of pollen and nectar. *A. mellifera* and *A. cerana* showed a special attraction to these flowers. Pollen foragers were more abundant. *P. rhoeas* (family Papaveraceae) blooms from February to April. Flowers are showy and rich with pollen. Pollen-laden foragers of *A. mellifera* can be observed during morning hours.

P. grandiflora (Portulacaceae) provides forage to honeybees during the summer dearth period under Punjab conditions and acts as subsistence flora, and profusely visited by *A. mellifera*, *A. cerana* and *A. florea*, *G. coronaria* (Asteraceae) blooms in February to April. These showy flowers provide pollen and nectar throughout the day, but peak foraging is noted at midday. Pollen foragers with full pollen loads in their corbiculae are pretty standard. *A. leptopus* (Polygonaceae), also called Mexican creeper or coral vine, has attractive bloom (Figure 2.27) of pink colour (white in some cultivars), which is in abundance during the summer dearth period (May–July) when honeybee colonies are in critical condition due to nutritional deficiency; thus, this vine acts as subsistence flora. Flowers provided both pollen and nectar to bee fauna.

FIGURE 2.23 *Apis florea* collecting nectar from *Spermadictyon suaveolens.*

Apis species frequently visited flowers (Figure 2.28) of *Rosa indica* (Rosaceae) and *Hamelia patens* (Rubiaceae) for nectar (Dalio, 2019a). Barring the previous ones, another ornamental flora related to bees is shown in Table 2.3.

Vegetables Crops as Bee Flora

Luffa aegyptiaca (Cucurbitaceae) flowering is an excellent source of nectar and pollen for *A. mellifera* (Figure 2.29). During the summer, dearth period, beekeepers have to feed the bees artificially. About 24.45- and 26.20-mg nectar is present in one male and female flower with a nectar sugar concentration of 24.82 and 25.75%. It also provides a sufficient amount of forage to wild honeybees like *A. florea* (Figure 2.30) and *A. cerana*.

According to an estimate, one feddan of luffa plantation can yield 35.44-kg nectar daily, which means 8.55 kg of honey. Flowers of other cucurbits like bottle gourd

FIGURE 2.24   *Apis mellifera* on *Dahlia* spp.

(*Lagenaria siceraria*), ridge gourd (*Luffa acutangula* L.), pumpkin (*Cucurbita maxima*) and *Cucurbita moschata* are a source of forage in summer. Pumpkin and luffa flowers provided a tremendous amount of nectar and pollen and frequently visited by honeybees, and more forage collection took place during morning hours (Dalio, 2019a).

Turnip (*Brassica rapa*; Brassicaceae) flowers are attractive to all types of honeybees (Figures 2.31 and 2.32) and provide both pollen as well as nectar. Flowers of other vegetables of the Brassicaceae family like *Brassica oleracea* var. Italica, *B. oleracea* var. capitata and *B. oleracea* var. botrytis (Figures 2.31–2.35) offered pollen as well as nectar during March and April in Punjab (India). *A. mellifera* is an efficient pollen gatherer while *A. cerana* collects nectar effectively, and side working or nectar robbing is standard in the case of *A. cerana* and *A. florea* (Figure 2.35).

Radish (*Raphanus sativus* L.; Brassicaceae) flowers are excellent nectar and pollen sources (Atwal, 2000; Abrol, 2010). Flowers opened in the morning and remained

FIGURE 2.25 *Apis mellifera* on *Papaver rhoeas.*

open for 2–3 days during March and April and showed strong attraction for honey bees (Figures 2.36 and 2.37). The flowers are actinomorphic with exposed nectaries and provide a landing platform of petals to bees to collect a sufficient amount of nectar and pollen. Due to such suitable structure and rewards, various insets, particularly honeybees, visit the bloom frequently. Pollen-laden foragers of *Apis* species are very common throughout the day.

Carrot (*Daucus carota*; Apiaceae) flowers are individually small and inconspicuous, but their aggregation to form compact umbels creates sufficient visual impact. Umbel functions as the primary unit to attract honeybees and other pollinators. Honeybees received pollen and nectar as reward for frequent visitation. *Apis* species (Figures 2.38 and 2.39) foraged on bloom from morning to evening during March and April, but peak foraging was observable from 1100 to 1300 hours (Dalio, 2019a).

Allium cepa (Amaryllidaceae) flowers are visited by honeybees mainly as a source of nectar (Dalio, 2019a). The percentage of nectar seekers was significantly more than

FIGURE 2.26 *Apis cerana* on *Papaver rhoeas.*

that of pollen gatherers (Kumar et al., 1985). Bees are attracted to the bloom due to more nectar with higher sugar concentration (Free, 1970). The concentration of sugar in onion nectar has been reported from 52 to 65% (Hagler et al., 1990) and from 70 to 75% (Akopyan, 1977). The nectaries are shallow, and if the nectar is not removed rapidly by insets, they can easily be seen glistening in the sunlight (Devi et al., 2014). Onion bloom (in March–April) provided pollen and nectar and visited by *Apis* species (Figure 2.40 and 2.41) more frequently in the late morning. Besides the previous ones, other vegetable crops are shown in Table 2.4.

Spice Crops as Bee Flora

The flowers of fennel (*Foeniculum vulgare* Mill; Umbelliferae) are umbelliferous and highly entomophilous. Honeybees visit its bloom (Figure 2.42) both for pollen and nectar from March to April. Colonies of *A. mellifera* flourish significantly by collecting

FIGURE 2.27   *Antigonon leptopus* flowering in full swing.

FIGURE 2.28   *Apis mellifera* foraging on rose.

TABLE 2.3

Some Other Ornamental Plants of Bee Interest

Sr. No.	Name of Plants	Family	Source	Blooming Month
1	*Ageratum conyzoides* L.	Compositae	P	July–September
2	*Althea rosea*	Malvaceae	N+P	April–May
3	*Antigonon leptopus*	Polygonaceae	N+P	May–July
4	*Aster* spp.	Compositae	N+P	July–September
5	*Bellis perennis*	Compositae	N+P	March–May
6	*Callistemon citrinus* Curt.	Myrtaceae	N+P	March–October
7	*Callistemon lanceolatus* D.C.	Myrtaceae	N	July–August
8	*Cassia fistula*	Fabaceae	N+P	May–June
9	*Centaurea cyanus* L.	Compositae	N+P	February–April
10	*Cosmos sulphureus*	Compositae	N+P	March–May
11	*Delphinium denudatum*	Ranunculaceae	P	March–April
12	*Delonix regia*	Bignoniaceae	N +P	March–May
13	*Gaillardia* spp.	Compositae	N+P	April–May
14	*Gladiolus* spp.	Iridaceae	N+P	May–July
15	*Ipomoea acuminata*	Convolvulaceae	N+P	Throughout the year
16	*Ipomoea palmata*	Convolvulaceae	N+P	July–August
17	*Jacaranda acutifolia*	Bignoniaceae	N+P	March–May
18	*Lagerstroemia* spp.	Lythraceae	N	May–September
19	*Lonicera sempervirens* L.	Caprifoliaceae	P	March–August
20	*Magnolia grandiflora*	Magnoliaceae	P	August–September
21	*Mesembryanthemum crystallinum*	Aizoaceae	N+P	February–March
22	*Narcissus poeticus*	Amaryllidaceae	N+P	February–March
23	*Quisqualis indica* L.	Combretaceae	N	May–September
24	*Ranunculus bulbosus*	Ranunculaceae	P	February–March
25	*Solidago longifolia*	Compositae	N+P	September–October
26	*Tecoma stans*	Bignoniaceae	N+P	February–April
27	*Thevetia peruviana*	Apocynaceae	N	April–June
28	*Zinnia* spp.	Compositae	P	July–October

Note: N= nectar, P=pollen, N+P=nectar and pollen.

pollen and nectar from fennel. Peak foraging activity has been reported (Narayana *et al.*, 1960; Baswana, 1984) from 1100 to 1400 hours because the highest amount of nectar and nectar sugar concentration was recorded between 1200 and 1400 hours (Atallah *et al.*, 1989). Coriander (*Coriandrum sativum* Linn; Umbelliferae) bloom has a pleasing fragrance due to nectar which attracts *Apis* species and other pollinators for foraging, thus helping in maintaining populations of bee colonies. They visit the flowering so abundantly that their humming can be heard on entering the field during the entire blooming season of the crop (February to April). The flowers are visited by bees more for nectar, but pollen is also collected in the absence of better pollen flora in the vicinity (Mishra, 1995). Maximum foraging took place in the afternoon. Honeybee species, including *A. mellifera* (Figure 2.43), *A. dorsata, A. cerana* and

FIGURE 2.29 *Apis mellifera* foraging on *Luffa aegyptiaca* flower.

FIGURE 2.30 *Apis florea* pollen gatherer on *Luffa aegyptiaca*.

FIGURE 2.31   *Apis mellifera* forager on turnip flowers.

FIGURE 2.32   *Apis florea* pollen collector on *Brassica rapa* flowers.

FIGURE 2.33 *Apis florea* nectar robbers on *Brassica oleracea*

A. florea, visited the bloom frequently (Painkra, 2019). *C. sativum* L. is a typical zygomorphic, and flowers have fleshy discs (stylopodium) which surround the ovaries from which nectar is secreted (McGregor, 1976). Funnel and coriander bloom was visited actively by dwarf, Indian and Italian honeybees.

Trachyspermum ammi (Apiaceae) is grown extensively in India. Its bloom has a solid and aromatic fragrance, which attracts honeybees to feed on nectar in March and April. *A. mellifera* (Figure 2.44), *A. cerana* and *A. florea* visit its flowering frequently. Unifloral raw honey of Bishop's weed has medicinal value and unique taste.

Black cumin (*Nigella sativa*) bloom is attractive to *A. mellifera*, and its foragers collect pollen and nectar from flowers. However, wild bees visit flowers with low frequency (Gary, 1979). The attractiveness of the Italian bee towards its bloom is due

FIGURE 2.34   *Apis florea B. oleracea var. capitata. var. botrytis* flower

to the availability of a higher volume of nectar with a characteristic flavour and more sugar concentration. Flower colour changes from a bud to a blooming stage, and *Apis* species, especially rock bee and Italian bee, visit the flowers. *A. mellifera* showed maximum foraging activities in the afternoon (1200–1400 hours) and considered a significant pollinator of the crop (Wahab Abd-El and Ebadah, 2011). The spice herb is reputed to be an excellent source of nectar for honeybees, particularly for *A. mellifera* (Amin, 1991).

Fenugreek (*Trigonella foenum-graecum*; Fabaceae) flowers are in full swing from February to April, which are dark yellow, and all types of the honeybees, especially *A. mellifera* and *A. florea* (Figures 2.45 and 2.46), visit more abundantly for nectar as well as pollen.

FIGURE 2.35   *Apis florea* visiting *Brassica oleracea* of *B. oleracea* var. italica. of *B. oleracea* var. capitata. var. botrytis flower.

Pollen-laden workers of *Apis* species can be seen throughout the day, but peak foraging activity can be noted during the late morning and early afternoon. *A. mellifera* bees were found foraging upto dusk. Nectar's aroma can be felt from a distance (Dalio, 2019a).

Anethum graveolens (Apiaceae) is an important spice crop as its leaves and seeds are used for flavouring food. It blooms from December to March and a good source of pollen and nectar for bees (Chaudhary, 2003). Two other spice crops attractive to honeybees are *Crocus sativus* L. (Iridaceae) and *Elettaria cardamomum* (Zingiberaceae). *C. sativus* L. provides enormous amount of nectar and pollen to honeybees from October to November (Atwal, 2000), but many cultivars show flowering during spring. *E. cardamomum* blossoms from July to November, which is a rich source of

FIGURE 2.36 *Apis mellifera* on radish flower.

nectar with good sugar concentration (37.48%) for *Apis* species (Sivaram, 2001), but a collection of both nectar, as well as pollen, has also been reported (Chaudhary and Rakesh, 2000). Bees forage during morning hours as anthesis takes place early in the morning.

Ocimum basilicum (Lamiaceae), also called royal herb or king of herbs, is well known for its sweet aroma and unique flavour. A study (Dalio, 2018a) showed that the bloom of the herb was rich with nectar and pollen and very attractive to honeybees (Figures 2.47–2.49). Flowers remain available almost around the year, thus helping to fill the gaps of floral scarcity. Foraging activity continued throughout the day, but maximum abundance was recorded during morning and afternoon in *A. cerana* and *A. mellifera*, respectively. Average abundance of *A. florea* foragers were very high (12 ± 2.03 bees/m^2/min) on sweet basil bloom to collect pollen and nectar (Dalio, 2018b).

FIGURE 2.37 *Apis cerana* on *Raphanus sativus* bloom.

Trees as Bee Flora

Eucalyptus spp. (Marantaceae) are evergreen trees of Australian origin but commonly found in India, which attain the gigantic size, and most of its species bloom in spring. These exquisite flowers are a reliable source of pollen and nectar for honeybees. Abundant nectar gets stored in a cup-shaped base of a flower. Traces of eucalyptol compounds are present in nectar, which provide unique flavour, aroma and appearance to its honey. Foraging activity of *A. mellifera* can be noted from dawn to dusk. However, both pollen and nectar are gathered, but nectar collection is more common. According to an estimate (Jhajj *et al.*, 1996), a fully grown *Eucalyptus* tree can produce 4 lakh flowers, yielding 15.5 kg of nectar

FIGURE 2.38 *Apis mellifera* forager on carrot flowering.

with a higher sugar concentration (54.7%), thus the high potential honey source for *Apis* species.

Dalbergia sissoo DC (Leguminosae), commonly known as North Indian rosewood, is an evergreen tree which blooms from March to May under Punjab (India) conditions. Flowers are pale yellowish, fragrant and having narrow tubular corolla, but honeybees can approach nectar easily. When these trees are present in large numbers near the apiary, they become primary bee flora that results in exceptional honey flow. Honey produced is very dense, has dark amber and more fructose content than glucose with solid flavour. Raw and monofloral honey produced by bees from rosewood bloom has medicinal properties to cure disorders of digestion, menstrual cycle and oral cavity.

FIGURE 2.39 *Apis florea* foraging on *Daucus carota* bloom.

Acacia species (Leguminosae) bloom is important bee forage, providing nectar as well as pollen. *Aconcinna* and *A. catechu* show flowering during March and April. *A. nilotica* has bright yellow flowers, blooms from October to December and March to April, while *A. juliflora* flowering occurs from March to May. These are excellent sources of nectar and pollen (Kumar and Sharma, 2016). In the case of *A. pycnantha*, flower heads and extrafloral nectaries are a source of food and readily visited by *A. mellifera* foragers (Giovanetti *et al*, 2015). Fresh pollen and nectar from *Acacia* dramatically raise the brood rearing activity. Another species, *A. ataxacantha*, is critical to apiculture in Nigeria, as beekeepers obtain vast amounts of honey during its blooming season. Bees visit the bloom throughout the day, but peak foraging activity is noted from 1000 to 1300 hours and seems to prefer flowers of this plant (Dukku, 2003).

FIGURE 2.40 *Apis mellifera* nectar seeker on onion bloom.

Toona ciliate (Meliaceae) is commonly grown in plains and lower hills of northern India as timber and avenue tree. Its flowers are small and cream-coloured, blooming from March to April, and are a source of abundant nectar and little pollen (Singh, 1948; Chaudhary, 1977). Observations (Mishra, 1995) revealed that nectar sugar concentration in freshly bloomed flowers was 26% which increased upto 72% in 48 hours in old flowers, nectar sugar value is 2.38 mg/flower/day and nectar is secreted for 4 days. Honey produced from its nectar is white to light amber in colour with a distinct flavour. The honey flow is short but effective.

Terminalia arjuna (Combretaceae) flowers are greenish or yellowish-white and sessile, which are in full swing during May and June, thus providing nectar to *A. Mellifera* and wild honeybees during the summer dearth period. In certain localities where plants are large, flowering offers sufficient nectar for bees to produce surplus honey (Singh, 1948) of amber to dark amber colour with solid flavour.

Bauhinia species (Leguminosae) bloom (Figure 2.50) is an excellent nectar source as it provides more volume of high-quality nectar to honeybees during morning hours

FIGURE 2.41 *Apis florea* foraging on *Allium cepa* flowering.

in October to December. Flowers are showy, attractive to honeybees and provide pollen along with nectar to various insect visitors, including *A. mellifera* (Figure 2.51). Trees other than the above-mentioned ones are enlisted in Table 2.5.

Weeds as Bee Flora

There are certain plants (however, minor bee flora) that bloom during the dearth period and provide floral rewards to honeybees facing harsh conditions and nutritional stress; such flora is called subsistence bee flora (Dalio, 2015a).

Cyperus rotundus (Cyperaceae) has a worldwide distribution in tropical and temperate regions, well known in more than 90 countries, a common perennial weed in Punjab, growing in crops and lawns, orchards and uncultivated lands. It flowers from May to October and an excellent pollen source for *A. mellifera*, which has been observed foraging during the early morning (540–900 hours). A higher abundance (35 bees/m²/min) was noted on its flowering during the summer dearth period at 700 hours. The foragers of Italian honeybees were found clinging to the spikelets, with pollen loads in their curbeculae (Dalio, 2018a).

Cannabis sativa (Cannabaceae) is a robust, tall-growing, herbaceous, dioecious and wild plant. Male plants are widespread as compared to female plants. This wild

TABLE 2.4

Some Other Vegetable Crops of Bee Interest

Sr. No.	Name	Family	Source	Months	References
1	*Abelmoschus esculentus* L	Malvaceae	N+P	April–August	Chaudhary (2003)
2	*Allium sativum* L.	Liliaceae	N+P	May–June	Singh (1962)
3	*Apium graveolens*	Umbelliferae	P	March–April	Abrol (2009)
4	*Asparagus officinalis*	Liliaceae	N+P	May–November	Abrol (2009)
5	*Citrullus fistulosus*	Cucurbitaceae	N+P	April–October	Abrol (2009)
6	*Citrullus lanatus*	Cucurbitaceae	N+P	April–June	Abrol (2009)
7	*Coccinia annuum* L.	Cucurbitaceae	N+P	June–August	Chaudhary (2003)
8	*Coccinia indica*	Cucurbitaceae	P	January–August	Chaudhary (2003)
9	*Cucumis melo* L.	Cucurbitaceae	N+P	April–June	Abrol (2009)
10	*Cucumis melo* L. var utilissimus Duth & Full	Cucurbitaceae	N+P	April–July	Chaudhary (2003)
11	*Cucumis sativus* L.	Cucurbitaceae	N+P	June–July September–October	Chaudhary (2003)
12	*Cucurbita maxima*	Cucurbitaceae	P	January–December	Sivaram (2001)
13	*Cucurbita moschata*	Cucurbitaceae	N+P	April–July	Abrol (2009)
14	*Cucurbita pepo*	Cucurbitaceae	N	April–June	Abrol (2009)
15	*Ipomea eriocarpa*	Covoluvaceae	N+P	December–January	Sivaram (2001)
16	*Lagenaria siceraria* (Molina) Standley	Cucurbitaceae	N+P	June–September	Chaudhary (2003)
17	*Lycopersicon esculentum*	Solanaceae	P	July–September	Sivaram (2001)
18	*Momordica charantia*	Cucurbitaceae	N+P	April–July	Abrol (2009)
19	*Sechium edule*	Cucurbitaceae	N+P	January–December	Sivaram (2001)
20	*Solanum melongena* L.	Solanaceae	P	March–September	Singh (1962)
21	*Trichosanthes anguina*	Cucurbitaceae	N+P	January–March	Sivaram (2001)

Note: N= nectar, P=pollen, N+P=nectar and pollen.

plant grows all over India but very common in Punjab in the non-cropland area (along the sides of roads, canals, railway tracks, wasteland etc.).

A. mellifera collects pollen from its bloom during morning and evening hours, but the significantly higher abundance of honeybees has been recorded during the morning compared to that in the evening. All the foraging bees were pollen gatherers, and no nectar foraging was observed. Flower sweeping activity and scrabbling behaviour of bees to collect pollen were seen during evening hours.

The weed is in full bloom during floral scarcity (May–July), thus helping to sustain bee colonies (Dalio, 2012) as pollen shortage is disastrous to bee colonies. Many other workers have also confirmed the pollen-collecting behaviour of honeybees from the bloom of this weed (Singh, 1962; Sharma and Raj, 1985; Brar *et al.*, 1989; Sharma and Gupta, 1993; Abrol, 1997).

FIGURE 2.42 *Apis mellifera* on fennel.

FIGURE 2.43   *Apis mellifera* on *Coriandrum sativum* bloom.

Trianthema portulacastrum (Aizoaceae) is an annual herb with prostrate, glabrous and a succulent stem with forked branches. It is the most common weed of Kharif crops, orchards and all vegetable crops grown during summer. It is abundantly found in cotton fields in the Malva region of Punjab (India), which produces and blooms with a rise of temperature (Walia and Uppal, 2001). *A. mellifera* bees forage on the bloom of this weed for pollen and nectar during morning (0730–1200) hours only. Average abundance of *A. mellifera* foragers was 5.89 bees/m²/min. It is good subsistence forage as it blooms abundantly during May and June, which is a lean period of the year (Dalio, 2015b).

Parthenium hysterophorus (Asteraceae) is a fast-spreading weed, which mainly grows on roadside, wastelands, orchards and now being introduced in field crops. It is an erect, many-branched, aromatic hairy herb with many-branched panicles

FIGURE 2.44   *Apis mellifera* on *Trachyspermum ammi.*

inflorescence having white flowers (Walia and Uppal, 2001). *A. mellifera* foragers visited the bloom during morning and evening. Maximum abundance
(13bees/m²/min) was observed from 800 to 845 hours. All the foraging bees were
pollen gatherers; on average, 5.28-mg pollen load per bee was recorded. *P. hys-*
terophorus is a minor but subsistent bee flora as it blooms during harassing summer
(May and June), which is a period of highly floral scarcity and significant bee flora
is absent. However, bees may neglect the weed when another bee flora is abundant
in the vicinity (Dalio, 2013).

Pollen collection by *A. melliferea* from *C. rotundus*, *T. portulacastrum* and *C.*
sativa was also confirmed by the author in another study (Dalio, 2018c).

Other weeds of bee interest (Dalio, 2015a) include *Tribulus terrestris*
(Zygophyllaceae), *Digera arvensis* (Amaranthaceae), *Cleome viscosa* L. (Cleomaceae)

FIGURE 2.45   *Apis florea* foraging on fenugreek flowers.

FIGURE 2.46   *Apis mellifera* pollen gatherer on fenugreek flowering.

FIGURE 2.47 *Apis cerana* on *Ocimum basilicum* bloom.

and *Cucumis trigonus* (Cucurbitaceae), *Ageratum conyzoides* (Compositae). *Taraxacum officinale* (Asteraceae) and *Stellaria media* (Caryophyllaceae) provide pollen and nectar to all types of honeybees (Figures 2.52–2.54) in the spring season. *D. arvensis* flowering (Figure 2.55) during August and September serves as minor bee flora. Some other weeds having status of bee flora are revealed in Table 2.6.

Floral Calendar

The flora of any area is characteristic of the natural setting and agro-climatic conditions. This varies from place to place. Before starting apiculture, knowing the availability of bee flora in the surrounding area is of utmost importance, as honeybees do not visit the bloom of all the plants because they have their floral preferences depending upon type, quality and quantity of floral rewards.

FIGURE 2.48   *Apis mellifera* on *Ocimum basilicum.*

To manage the stationary as well as migratory beekeeping in a rational way, the floral calendar is an essential tool, which is a timetable to guide beekeepers regarding the approximate date and duration of blossoming periods of plants of bee interest. Preparation of a floral calendar requires data collection on initiation and cessation of blooming of bee flora and foraging behaviour of *Apis* species (nectar and or pollen gatherers) at various day hours throughout the year within flight range of apiaries. Such details indicate honey flow and lean periods of the year, which helps manage the bees accordingly. To ensure the accuracy of the floral calendar, careful scientific field observations of foraging responses of honeybees to flowers of various plants are necessary. Pollen and nectar yield of a particular plant species may vary in different climatic conditions. It may be a significant bee flora in one region and minor in the other. The status of plants as bee flora is determined by the relative abundance and foraging behaviour of honeybees on their flowers. Microscopical observations of pollen and pollen analysis of honey stored in bee colonies are also reliable methods of finding the exact plant source.

FIGURE 2.49 *Apis florea* on *Ocimum basilicum.*

Theodor of nectar or honey also helps to confirm plant source. Adding more and more bee flora to any area and diversification of crops will help increase its bee carrying capacity.

Why do Honeybees Make Honey?

The priority of honeybee colonies is long-term existence. In warm months of the year, abundant bee flora is available; therefore, they collect nectar, convert it into honey and store it as reserve food (because its fermentation does not take place) to survive the winter. During winter, working hours decrease and bee activity reduce

FIGURE 2.50   *Bauhinia variegata* flowering in full swing.

FIGURE 2.51   *Apis mellifera* foraging on *Bauhinia variegata* flower.

TABLE 2.5

Some Other Trees of Bee Interest

Sr. No.	Name of Tree	Family	Source	Period of Blooming	References
1	*Aegle marmelos* L. Correa	Rutacea	N+P	July–August	Abrol (2009)
2	*Albizia chinensis* (Osb.) Merr	Leguminosae	N+P	April–May	Abrol (2009)
3	*Albizia lebbeck* Linn.	Leguminosae	N+P	April–May	Sharma and Raj (1985)
4	*Azadirachta indica* A	Meliaceae	N+P	March–April	Atwal (2000)
5	*Bombax ceiba* L.	Bombacaceae	N+P	January–March	Naim and Phadke (1976)
6	*Butea monosperma* (Lamk.)	Papilionaceae	N+P	January–February	Khan (1948)
7	*Cocos nucifera* L.	Arecaceae	N+P	January–December	Anonymous (1959)
8	*Cordia dichotoma*	Boraginaceae	N+P	April–May	Abrol (2009)
9	*Grevillea robusta* A.	Proteaceae	N+P	April–June	Atwal (2000)
10	*Hevea brasiliensis*	Euphorbiaceae	N	March	Deodasan (1972)
11	*Lagerstroemia entrada*	Lythraceae	N+P	July–September	Diwan and Rao (1972)
12	*Lagerstroemia speciosa*	Lythraceae	N+P	April–July	Abrol (1997)
13	*Madhuca longifolia* (Koenig) J.F.	Sapotaceae	N+P	February–March	Naim and Phadke (1976)
14	*Melia azedarach*	Meliaceae	N	March–April	Abrol (2009)
15	*Mitragyna parvifolia*	Rubiaceae	P	January–August	Abrol (2009)
16	*Moringa pterygosperma* Gaertn.	Moringaceae	N+P	February–April	Nair and Singh (1974)
17	*Morus alba*	Moraceae	P	February–March	Abrol (2009)
18	*Robinia pseudoacacia*	Leguminosae	N	May–June	Giovanetti (2019)
19	*Salmalia malabarica*	Bombacaceae	N	February–March	Singh (1962)
20	*Sapindus detergens* Roxb.	Spindaceae	N+P	May	Sharma and Raj (1985)
21	*Sapindus emarginatus* Vahl.	Spindaceae	N+P	October–November	Krishnaswamy (1970)
22	*Sapindus mukorossi* Gaertn	Spindaceae	N+P	May	Atwal (2000)
23	*Terminalia bellirica* (Gaertn.) Roxb.	Combretaceae	N+P	April–May	Anonymous (1959)
24	*Tamarindus indica*	Leguminosae	N	April–July	Singh (1962)
25	*Terminaliachebula* Retz.	Combretaceae	N	April–May	Anonymous (1959)

Note: N= nectar, P=pollen, N+P=nectar and pollen.

FIGURE 2.52   *Apis mellifera* on *Ageratum conyzoides.*

considerably, and floral scarcity prevails (winter dearth period). All the worker bees of the colony play a significant role in preparing and storing honey that will ultimately help to survive the settlement for a long time despite floral deficiency and adverse weather conditions. Stored honey in honeybee colonies is food security for the future.

Mechanism of Honey Production

Bees are an integral part of our ecosystems by converting nectar into honey in a complicated and efficient way like excellent chemists, a supersaturated solution made by providing heat, enzymes or other biochemicals. Honey making is a complex and systematic process, which results from the coordination of cooperative activities of the

FIGURE 2.53 *Apis mellifera* on *Taraxacum officinale.*

field going and the housekeeping members of the colony. Forager bees bring nectar or honeydew from the crops into the hive, and they consume a tiny part of it during collection while most of it is used for honey production. The brought material is not stored in comb cells directly.

But bees transfer the nectar to other bees. During this process, many physical and chemical changes occur like various types of sugars get converted into monosaccharides. This raw material is stored in comb cells, and water starts evaporated, which is speeded up by repeated relocation and fanning behaviour.

Honey is flower nectar collected, regurgitated after adding enzymes and dehydrated to enhance its nutritional properties. Nectar has about 80% water and cannot be stored by honeybees for a long time as its fermentation occurs. Therefore, they convert it into honey which can be stored as reserve food for a long time. Sometimes bees collect honeydew secretion of aphids or other plant-sucking insects instead of nectar. Honey formed from the collection of honeydew is not of good quality. The sweetness of such honey is much lesser than nectar honey and may cause complications in bee colonies. It is commonly called honeydew honey.

FIGURE 2.54 *Apis mellifera* on *Stellaria media.*

FIGURE 2.55 *Digera arvensis* bloom in full swing.

TABLE 2.6

Some Other Weeds of Bee Interest

Sr. No.	Name of the Weed	Family	Source	Blooming Time
1	*Ageratum conyzoides* L.	Compositae	N+P	January–June
2	*Anagallis arvensis*	Primulaceae	N	December–March
3	*Anemone vitifolia* Buch Ham.	Ranuculaceae	P	Mid-May–September
4	*Argemone mexicana* L.	Papaveraceae	P	March–July
5	*Calotropis procera*	Asclepiadaceae	N	June–September
6	*Carthamus oxyacantha* Bieb.	Asteraceae	N+P	May–June
7	*Chenopodium album*	Chenopodiaceae	P	March–May August–September
8	*Chenopodium ambrosioides* L.	Chenopodiaceae	P	March–May August–September
9	*Chionachne koenigii* Thw.	Gramineae	N+P	July–mid-August
10	*Cichorium intybus* L.	Compositae	N	March–May
11	*Cleome viscosa* L.	Capparidaceae	N+P	Mid-June–September
12	*Convolvulus arvensis* L.	Convolvulaceae	N+P	January–December
13	*Datura* spp.	Solanaceae	P	June–October
14	*Lagascea mollis* Cav.	Compositae	N	November
15	*Polygonum* spp.	Polygonaceae	N+P	July–October
16	*Rumex maritimus*	Polygonaceae	P	February–March
17	*Stellaria media* (L.) Vill.	Caryophyllaceae	N+P	January–April
18	*Taraxacum officinale* Weber	Compositae	N+P	March–November
19	*Trifolium repens* L.	Leguminosae	N+P	March–August

Note: N= nectar, P=pollen, N+P=nectar and pollen.

Nectar Secretion and its Collection

Nectar production is an evolutionary adaptation that attracts insects to fulfil the pollination requirements of the plant concerned. Nectar is a sugar-rich liquid secreted by a glandular portion of flowers called nectaries.

These floral nectaries are present within the flower, which provide nectar as a reward to pollinators. However, such glandular portions may be extrafloral, also called extranuptial nectaries (extrafloral nectaries). Extrafloral nectar though small in quantity but is greedily sought by insects (Darwin, 1859). Nectar collection from extrafloral nectaries on leaves and stems of plants and displaying honeydew from aphids and leafhoppers has also been reported in the literature (Atwal, 2000). Sight and smell enable the bees to locate flowers, and they are probably guided to nectaries by the difference in the odour of the nectaries and the rest of the flower. Large loads of nectar of about 70 mg in Italian bee had been recorded (Park, 1922, 1925). However, average loads of nectar noted during favourable conditions were 30–40 mg. Nectar collection occurs with proboscis and stored in their honey stomach or crop that can hold 70 mg of nectar. The bee visits upto 1500 flowers to fill its stomach; however, this number varies considerably according to species of blooming crop. To prepare one pound of honey, bees must visit about 2 million flowers and about 100 flowers in one foraging trip. Older (Scout) bees help find forage and communicate with hive bees about food resource and may travel upto 5 mi in search of forage.

Unloading Nectar

The field bees bring nectar in their honey vesicle into the hive, which is not stored in comb cells directly but given to hive-bee or bees waiting to take the load. If the field bee is returning from a rich source, it may perform a communication dance, giving a taste of forage to other workers. Bees, which transfer the nectar to other bees, stay face to face, touching by their snouts. During nectar distribution, the field bee opens its mandibles apart and brings a drop of nectar, and the house bee protrudes out its tongue and sucks this nectar. Then they transfer the nectar to other bees. Each bee goes on adding enzymes to convert complex sugars of dilute and unripe honey into monosaccharide sugars. Mouth-to-mouth give and take help in condensing it through evaporation, which also continues when unripe nectar is poured in comb cells and further accelerated by relocation did several times.

Special Bee Behaviour for Honey-Making

When honey flora is abundant and nectar is being brought rapidly, the housekeeping bees of the colony store this invertase-containing nectar temporarily on the upper walls of comb cells in the form of hanging droplets. The spherical surface increases surface area. Thus, the evaporation rate of water increases considerably. After sufficient concentration is attained, these droplets are again sucked to fill their stomach and start the ripening process in a suitable comb area. A droplet of a mixture of nectar and enzymes is brought out of the pre-oral cavity, holding it between the palpi and distal proboscis of the tongue, which increases to maximum volume and then reduced in size by sucking it inwards. This activity takes 5–10 seconds. Pollen grains from the wort are also gathered and sent to the stomach; thus, pollen grains are reduced. This procedure of taking and bringing out droplets is repeated for about 20 minutes. However, the above-mentioned intervals of time may vary under different conditions; when droplet attained a suitable concentration level; bee enters in comb cell and paints the upper wall of it with that liquid with the help of mandibles by moving its head from side to side. If unripe honey is already present in the cell, it adds droplet without painting, but sugar conversion into simple sugars takes many hours. Evaporation of water from this unripe honey continues for 2–3 days. The fanning behaviour of worker bees helps a lot to speed up the evaporation by circulating air throughout the hive. During fanning behaviour, bees fan their wings very rapidly (11,000 times/min) and vigorously reduce the moisture content of unripe honey. The bees continue to shift unripe honey (Figure 2.56) from one cell to another, which also help to reduce water content. Eventually, the honey will have a moisture content of about 18%, and it is stored on the top of the comb and sealed with wax caps (Figure 2.57)

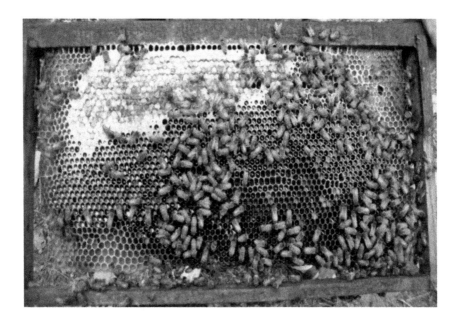

FIGURE 2.56   Comb with unripe and ripened honey.

FIGURE 2.57   Ripened and sealed honey on the top of the comb.

Role of Enzymes and Floral Ingredients

Honeybees not only collect, regurgitate and store nectar but also change it chemically and physically. A critical enzyme secreted by the salivary glands is added to the collected nectar in the honey stomach, which helps break sucrose into glucose and fructose (monosaccharides or simple sugar). It is the beginning of honey formation. Fructose is sweeter as compared to glucose, thus helps to make honey sweeter. Amylase is used to break down amylose into glucose which also adds sweetness to honey. Another enzyme, glucose oxidase, which honeybees synthesise and add into honey, catalyses glucose oxidation to hydrogen peroxides and gluconolactone. Its reaction with nectaries responsible for the by-products of gluconic acid and hydrogen peroxides contributes to acidity, antibacterial properties and the stabilisation of pH of the honey.

Catalase helps to change hydrogen peroxide into water and oxygen to keep its content low. It is of plant origin, and its concentration depends on the number of pollen grains in honey and varies considerably in love from different floral sources. There is high content (50–85% of total amino acids) of proline, which plays an essential role in mixing nectar solution with enzyme solutions. Fructose/glucose ratio, granulation, aroma, flavour, pollen content (protein content), ash content, flavonoids and volatile phenolic substances vary with floral resources variation. When bees collect nectar from different flowers, such honey is called polyfloral or multivariate or multifloral like wildflower honey. If nectar is collected from flowers of one species, love formed is called univarietal or monofloral. Such honey has characteristic flavour, aroma and colour.

Conclusion and Future Strategies

Flowers are the mainstay of a bee's life. Flowering plants (Angiosperms) and honeybees have a mutualistic relationship as both of them get the benefit; bees get food in the form of nectar and pollen from flowering plants and facilitate their pollination. Honeybees visit a variety of plants called bee flora to collect nectar and pollen. The qualities of good bee flora for commercial apiculture include long blooming period, higher flower density, the nectar with higher sugar concentration, and easily accessible rewards in flowers and availability in abundance in the areas around the apiaries. Availability of more pollen results in more brood rearing, which leads to robust colony build-up. Pollen sources are essential for full exploitation of honey flow, as a force of foragers is required to collect large honey stores. Nectar is a mostly mixture of carbohydrates that provides energy for flight, foraging, hive activities and brood-rearing etc., as well as a raw material for honey production. Honey is a nutritious, delicious and viscous hive product produced amazingly by the teamwork of field and housekeeping members of honeybee colonies. This complicated process includes a collection of floral nectar, the addition of enzymes and its dehydration by repeated relocation, heat and fanning behaviour.

The potential of bee flora of any ecosystem depends upon the density, distribution and blossoming of bee plants of different species existing in the area. Gathering knowledge about such flora, including identification, floral biology and degree of attractiveness for honeybees, is an integral part of beekeeping practices. Floral needs may be lesser for small-scale beekeeping, but to promote it as an employment generating activity and to increase yields of entomophilous crops through pollination that is directly linked to food security, more bee colonies are required. To support such a vast number of settlements, the management of bee pasture is essential. During lean periods, sugar syrup and pollen substitutes are fed to the domesticated colonies, but an artificial diet cannot fulfil all nutritional requirements. The time of absence and honey flow periods varies from area to area, so migratory beekeeping is recommended to ensure a continuous supply of natural forage.

Lacking information about beneficial plants for bees is a significant limitation for apicultural development. Attempts should be directed to find out or develop plants having more attractive and helpful blooms for honeybees. Essential factors for bee attractiveness or foraging cues like nectar, aroma, colour and flower morphometric should be studied and incorporated into new varieties/cultivars of different crops. The selection of plant stock based on honeybee foraging behaviour is very important. Pollen and nectar yield, as well as their nutritional quality, should be essential consideration while developing new cultivars. More research work may be done to know what makes pollen attractive to honeybees, so chemicals involved can be sprayed on desired crops for alluring bee foragers to collect more pollen.

For more successful beekeeping, apiculturists should have a thorough knowledge of the floral calendar, foraging behaviour of honeybees and management of bee flora. Efforts to make floral calendars of all states or regions should be done, and such exercise will provide guidelines for migratory beekeeping. More research on nectariferous and polleniferous plants and propagation of evaluated flora, diversification of crops with a focus on apiculture and growing trees of bee interest may increase the beekeeping potential of the area for fruitful beekeeping. Beekeepers, farmers, extension workers, scientists working on honeybees, plant breeders and environmentalists will have to work together to manage bee forage for developing apiculture and conservation of wild honeybees.

REFERENCES

Abrol, D.P. 1997. Food of honeybees, bee flora and honey flow periods. in D.P. Abrol. Ed., Bees and beekeeping in India. Kalyani Publishers, Ludhiana, India. pp. 108–151.

Abrol, D.P. 2009. Bees and beekeeping in India. Kalyani Publishers, New Delhi. pp. 125–162.

Abrol, D.P. 2010. Beekeeping – A comprehensive guide on bees and beekeeping. Scientific Publishers, Jodhpur. p. 896.

Akopyan, G.A. 1977. Pollination of onion seed plants. *Biol. Zh. Arm.* 20 (7): 88–89.

Amin, G.R. 1991. Popular medicinal plants of Iran. Iranian Research Institute of Medicinal Plants, Tehran, 1–66. (In Persian)

Anonymous. 1959. Bee flora of Karnataka and Kerala. *Indian Bee J.* 21 (7/8): 90–92.

Atallah, M.A., F.K. Aly and H.M. Eshbah. 1989. Pollen gathering activity of worker honeybee on field crops and medicinal plants in Minia region, Middle Egypt. *Proceedings of the Fourth International Conference on apiculture in Tropical Climates*, Cairo, Egypt. November 6–10, 1988 pp. 109–115.

Atwal, A.S. 2000. Essentials of beekeeping and pollination. Kalyani Publishers, New Delhi. pp. 198–218.

Atwal, A.S., S.S. Bains and B. Singh. 1970. Bee flora for four species of *Apis* at Ludhiana. *Indian J. Ent.* 32 (4): 330–334.

Baswana, K.S. 1984. Role of insect pollination on seed production in Coriander and Fennel. *South India Hort.* 32: 117–118.

Brar, H.S., G.S. Gatoria and B.S. Chahal. 1989. Bee flora of Punjab, its relative utility and Calendar of availability of honeybees. *Indian J. Ecol.* 16 (2): 159–163.

Chaubal, P.D. and S.Y. Kotmire. 1980. Floral Calendar of bee's forage plants of Sagarmal (India) *Indian Bee J.* 42 (3): 65–68.

Chaudhary, R.K. 1977. Bee forage in Punjab Plains, India, Pathankot and adjacent villages. *Indian Bee J.* 39 (1/4): 5–20.

Chaudhary, O.P. 2003. Evaluation of honeybee flora of the north eastern region of Haryana. *J. Palynol.* 39: 127–141.

Chaudhary, O.P. and K. Rakesh. 2000. Studies on honeybee foraging and pollination in cardamom (*Elettaria cordamomum* Maton). *J. Spices Aromatic Crops* 9 (1): 37–42.

Conner, J.K. and S. Rush. 1996. Effect of flower size and number on pollinator visitation to wild radish. *Raphanus raphanistrum Oecologia* 104: 234–245.

Dalio, J.S. 2003. Relative pollinating efficiency of *Apis* species on *Brassica napus* L. Ph.D. thesis submitted to Jiwaji University Gwalior, India.

Dalio, J.S. 2012. *Cannabis sativa* – An important subsistence pollen source for *Apis mellifera. IOSR J. Pharm. Biol. Sci.* 1(4): 1–3.

Dalio, J.S. 2013. Foraging activity of *Apis mellifera* on *Parthenium hysterophorus. IOSR J. Pharm. Biol. Sci.* 7(5): 1–4.

Dalio, J.S. 2015a. Status and problems of beekeeping in Mansa district of Punjab. *IOSR J. Pharm. Biol. Sci.* 10(2): 8–12.

Dalio, J.S. 2015b. Foraging behaviour of *Apis mellifera* on *Trianthema portulacastrum. J. Entomol. Zool. Stud.* 3(2): 105–108.

Dalio, J.S. 2017a. Foraging behaviour of honeybees on *Jatropha integerrima. J. Res.* 8(1): 98–101.

Dalio, J.S. 2017c. Strategic foraging of *Apis mellifera* on *Jasminum grandiflorum flowers. Proceedings of National conference*, Arni University, India (ISBN 978-93-5268-453-3), pp. 170–174.

Dalio, J.S. 2017d. *Combretum Indicum*: an excellent floral resource for *Apis* species. *Proceedings of National Conference*, Arni University, India (ISBN 978-93-5268-453-3), pp. 175–181.

Dalio, J.S. 2017e. *Cestrum nocturnum*: an excellent bee flora. *Proceedings of National Conference*, Arni University, India (ISBN 978-93-5268-453-3), pp. 182–188.

Dalio, J.S. 2017f. *Campsis tagliabuana*: an excellent floral resource for *Apis mellifera. Proceedings of National Conference*, Arni University, India (ISBN 978-93-5268-453-3), pp. 207–214.

Dalio, J.S. 2018a. Status of sweet basil as bee flora. *Int. J. Rec. Sci. Res.* 9(1) 23715–23717.

Dalio, J.S. 2018b. Foraging activity of dwarf honeybee (*A. florea*) on bloom of *Ocimum basilicum* L. *J. Res. Agric. Anim. Sci.* 5(1): 11–14.

Dalio, J.S. 2018c. Pollen collecting activity of *Apis mellifera. J. Entomol. Zool. Stud.* 6(2): 352–355.

Dalio, J.S. 2018d. Foraging behavior of *Apis* species on the bloom of *Ziziphus mauritiana* L. *Int. J. Appl. Res.* 4: 53–57.

Dalio, J.S. 2019a. Ability of different *Apis* species to explore and manipulate various floral resources. *Proceedings of National Conference*. Career Point University Hamirpur (ISBN 978-93-5391-471-4), pp. 132–138.

Dalio, J.S. 2021a. Insect fauna visiting *Tecoma castanifolia*. in Frontiers in science and technology, Grin Verlag, Germany (ISBN 9783346354082), pp. 93–102.

Dalio, J.S. 2021b. *Apis mellifera* and other insect fauna visiting *Spermadictyon suaveolens* Roxb bloom. In Frontiers in science and technology. Grin Verlag, Germany (ISBN 9783346354082), pp. 103–110.

Darwin, C. 1859. On the origin of species by means of natural selection, or the preservation of favoured races in the struggle for life. London: John Murray, (Google Scholar). http://scholar.google.com/scholar_lookup?title=On%20the%20action%20of%20sea-water%20on%20the%20germination%20of%20seeds&author=C.%20Darwin&journal=Journal%20of%20Proceedings%20of%20the%20Linnean%20Society%20of%20London%20%28Botany%29&volume=1&pages=130-140&publication_year=1857.

Deodasan, A. 1972. Rubber plantation and beekeeping. *Indian Bee J*. 34(1/2): 38–39.

Deodikar, G.B., C.V. Thakar and R.P. Phadke. 1957. High nectar Concentration in floral nectaries of silver oak *Grevillea robusta*. *Indian Bee. J*. 19(7/8): 84–85.

Devi, S., R. Gulati, K. Tehri and Asha. 2014. Diversity and abundance of insect pollinators on *Allium cepa* L. *J. Entomol. Zool. Stud*. 2(6): 14–18.

Diwan, V.V. and G.V. Rao. 1972. Altitudinal variation in bee forage of west Coorg. *Indian Bee. J*. 33(3): 39–50.

Doull, K.M. 1966. The relative attractiveness to pollen collecting honeybees of some different pollen. *J. Apicult. Res*. 5(1): 9–13.

Dukku, U.H. 2003. *Acacia ataxacantha*: a nectar plant for honeybees between two dearth periods in the Sudan savanna of northern Nigeria. *Bee World* 84(1): 32–34.

Elle, E. and R. Carney. 2003. Reproductive assurance varies with flower size in *Collinsia parviflora* (Scrophulariaceae). *Am. J. Bot*. 90: 888–896.

EL-Kazafy, A. Taha. 2007. Importance of Banana *Musa* Sp. (*Musaceae*) for honeybee *Apis mellifera* L. (Hymenoptera: Apidae) in Egypt. *Bull. Ent. Soc. Egypt*. II: 125–133.

Free, J.B. 1970. Insect pollination of crops. Academic Press, London.

Free, J.B. 1993. Insect pollination of crops. 2nd ed. Academic Press, London. pp. 06–104.

Gary, N.E. 1979. Factors that affect the distribution of foraging honeybees. *Proceedings of IVth International Symposium on pollination*. pp. 53–356.

Giovanetti, M. 2019. Foraging choices balanced between resource abundance and handling concerns: how the honeybee, *Apis mellifera*, select the flowers of *Robinia pseudoacacia*. *Bull. Entomol. Res*. 109(3): 316–324.

Giovanetti, M., M.M. Lippi, B. Foggi and C. Giuliani. 2015. Exploitation of invasive *Acacia pycnantha* pollen and nectar resources by the native bee *Apis mellifera*. *Ecol. Res*. 30: 1065–1072.

Goulson, D. 1999. Foraging strategies of insects for gathering nectar and pollen, and implications for plant ecology and evolution. *Evol. Syst*. 2/2: 185–209.

Goulson, D., J.W. Chapman and W.O.H. Hughes. 2001. Discrimination of unrewarding flowers by bees: direct detection of rewards and use of repellent, scent marks. *J. Insect. Behav*. 14: 669–678.

Gupta, J.K., R.K. Thakur and J. Kumar. 1986. Nectar of wild olive, *Elaeagnus umbellata* Thunb and foraging intensity of different insect visitors. *Indian Bee J*. 48: 40–42.

Gupta, J.K., J. Kumar and R. Kumar. 1990. Observations on the foraging activity of honeybees and nectgar Toon, Toona ciliate Roem. *Indian Bee J*. 52: 34–35.

Hagler, J.R., A.C. Cohen and G.M. Loper. 1990. Production and composition of onion nectar and honeybee (hymenoptera: Apidae) foraging activity in Arizona. *Environ. Entomol.* 19(2): 327–331.

Heinrich, B. 1979. Resource heterogeneity and pattern of movement in foraging bumble bees. *Oecologia* 40: 235–245.

Hooper, S.D. 2009. OCBIL theory: towards an integrated understanding of the evolution, ecology and conservation of biodiversity on old, climatically buffered, infertile landscapes. *Plant Soil.* 332: 49–86.

Inouye, D.W. 1983. The ecology of nectar robbing. The biology of nectaries. Columbia University Press, New York, NY.

Jhajj, H.S., G.S. Gatoria and D.R.C. Bakhetia. 1996. Three decades of beekeeping research in the Punjab. National Agricultural Technology Information Centre, Ludhiana (India). p. 21.

Kato, M. 1988. Bumble bee visits to *Impatiens* spp.: Pattern and efficiency. *Oecologia* 76: 364–370.

Kearns, C.A. and D.W. Inouye. 1993. Techniques for the pollination biologist. University Press of Colorado, Boulder, CO.

Khan, M.S.A. 1948. Some important nectariferous plants and pollen sources of Bhopal State, central India. *Indian Bee J.* 10(11/12): 107–108.

Kohli, N. 1958/59. Bee flora of northern India. *Indian Bee J.* 20: 113–118, 132–134, 150–151, 178–179, 192–193, 21: 7–8, 31–32, 61–62, 83–85, 106–107, 127–128.

Krishnaswamy, S.V. 1970. Soap nut trees: a nectar source. *Indian Bee J.* 32(3/4): 83.

Kudo, G. and L.D. Harder. 2005. Floral and inflorescence effects on variation in pollen removal and seed production among six legume species. *Funct. Ecol.* 19: 245–254.

Kumar J., R.C. Mishra and J.K. Gupta. 1985. The effect of mode of pollination on *Allium* species with observations on insects as pollinators. *J. Apic. Res.* 24(1): 62–66.

Kumar, D and V. Sharma. 2016. Evaluation of *Acacia* species as honeybee forage potential. *Int. J. Sci. Res.* 5(1): 1726–1727.

Leiss, K.A., K. Vrieling and P.G. Klinkhamer. 2004. Heritability of nectar production in *Echium vulgare. Heredity (Edinb).* 92(5): 446–451.

Levin, M.D. and G.E. Bohart. 1955. Selection of pollen by honeybees. *Am. Bee J.* 95(10): 392–393.

Marden, J.H. 1984. Remote perception of floral nectar by bumble bee. *Oecologia* 74: 232–240.

Macukanovic-Jocic, M. and L. Djurdjevie. 2005. Influence of microclimatic conditions on nectar exudation in *Glechomahirsuta* W.K. *Arch. Biol. Sci. Belgrade,* 57(2): 119–126.

McGregor, S.E. 1976. Insect pollination of cultivated crop Plants. USDA Handbook, Washington, DC.

Mishra, R.C. 1995. Honeybees and their management in India. ICAR Publication, New Delhi. p. 168.

Mishra, R.C., J.K. Gupta and J. Kumar. 1985. Nectar sugar production in different cultivars of peach, *Prunus persica* L. *Indian Bee J.* 47: 37–38.

Mishra, R.C. and S.K. Sharma. 1989. Growth regulators affect nectar pollen production and insect foraging in *Brassica* seed crop. *Curr. Sci.* 57: 1297–1299.

Moller, A.P. and G. Sorci. 1998. Insect preference for symmetrical artificial flowers. *Oecologia* 114: 37–42.

Naim, M. and K.G. Phadke. 1976. Bee flora and seasonal activity of *Apis cerana indica* at Pusa (Bihar). *Indian Bee J.* 38(1/4): 13–19.

Narayana, E.S., P.L. Sharma and K.G. Phadke. 1960. Studies on requirement of various crops for insect pollinators. I. insect pollinators of saunf (*Foeniculum vulgare*) with particular reference to the honeybee at Pusa (Bihar). *Indian Bee J.* 22(1/3): 7–11.

Nair, P.K.K. and K.N. Singh. 1974. A study of two honey plants, *Antigonon leptopus* Hook and *Moringa pterigosperma* Gaer. *Indian J. Hort.* 31: 375–379.

Park, O.W. 1922. Time and labor factors in honey and pollen gathering. *Am. Bee J.* 6: 254–255.

Park, O.W. 1925. The minimum flying weight of the honeybee. *Iowa St. Apiarist Rep.* 63: 83–90.

Painkra, G.P. 2019. Foraging behavior of honeybees on coriander (*Coriandrum sativum* L.) flowers in Ambikapur of Chhattisgarh. *J. Entomol. Zool. Stud.* 7(1): 548–550.

Percival, M. 1946. Observations on the flowering and nectar secretion of *Rubus fructicosus*. *New Phytol.* 45: 111–123.

Percival, M.S. 1965. Floral biology. Pergamon Press, Oxford. 243 pp.

Phillips, R.D., S.D. Hopper and W.D. Kingsley. 2010. Pollination ecology and possible impacts of environmental change in the Southwest Australian Biodiversity Hotspot. *Philos. Trans. Royal Soc. B.* 365: 517–528.

Saraf, S.K. 1973. Honey flow in Kashmir Valley. *Indian Bee J.* 35(1/4): 50–51.

Sharma, O.P. and D. Raj. 1985. Diversity of bee flora in Kangra, Shivaliks and its impact on beekeeping. *Indian Bee J.* 47: 21–24.

Sharma, H.K. and J.K. Gupta. 1993. Diversity and density of bee flora of Solan region of Himachal Pradesh (India). *Indian Bee J.* 55(1/2): 9–20.

Shuel, R.W. 1992. The production of nectar and pollen. in: Graham, J.M. Ed., The hive and the honeybee. Bookcrafters, Chelsea, MI. pp. 401–436.

Singh, S. 1948. Some important honey plants of the Punjab (India). *Rep. Lowa State Apiar.* 34–42.

Singh, S. 1962. Beekeeping in India. ICAR, New Delhi. pp. 92–123.

Singh, B., M. Kumar, A.K. Sharma and L.P. Yadav. 2006. Relative abundance of insect visitors on litchi (*Litchi chinensis* Sonn.) bloom. *Environ. Ecol.* 24: 275–277.

Sivaram, V. 2001. Honeybee flora and beekeeping in Karnataka State, India. *Proceedings of the 37th International Apicultural Congress*. Durban, South Africa, *APIMONDIA* (ISBN 0-620-277 68-8).

Srawan, B.S. and B.S. Sohi. 1985. Phyto sociological studies on *Apis mellifera* L. and *Apis cerana* F. in Punjab, India. *Indian Bee J.* 47: 15–18.

Taha, R.A., W. Badawy and E.-K. Taha. 2018. Sesame, *Sesamum indicum* L. (Pedaliaceae) Foraging behavior of honeybees, *Apis mellifera* L. and physio-chemical properties of honey. *J. Plant Protect. Pathol.* 9(8): 483–488.

Vansell, G.H., W.G. Watkins and R. K. Bishop. 1942. Orange nectar and pollen in relation to bee activity. *J. Econ. Entomol.* 35: 321–323.

Wahab Abd-El, T.E. and I.M.A. Ebadah. 2011. Impact of honeybee and other insect pollinators on the seed setting and yield production of black cumin, *Nigella sativa*. L. *J. Basic Apicul. Sci. Res.* 1(7): 622–626.

Walia, U.S. and R.S. Uppal. 2001. Weeds. in Singh, A.P. Ed., Field problems of fruit crops, Foll Printer, Ludhiana (India). pp. 188–239, 192.

Waser, N.M. 1983. The adaptive nature of floral traits: ideas and evidence. in Real, L. Ed., Pollination biology. Academic Press, New York, NY. pp. 241–285.

Wesselingh, R.A. and M.L. Arnold. 2000. Nectar production in *Louisiana iris* hybrids. *Int. J. Plant Sci.* 161(2): 245–251.

Wetherwax, P.B. 1986. Why do honeybees reject certain flowers? *Oecologia* 69: 567–576.

3

Quality Honey Production, Processing, and Various Mechanisms for Detection of Adulteration

Sunita Saklani
Govt Degree College, Sujanpur Tihra, India

Nitesh Kumar
Himachal Pradesh University, Shimla, India

CONTENTS

DOI: 10.1201/9781003175964-3

Introduction

A quality honey is an organic product of nature, which is prepared inside the hives by honeybees and is not altered or blended with other substances. A quality honey is well illustrated by various religious scripture: as in ancient Indian Vedas, it is described as 'Sarv Vai Madhu Anantam Vai Madhu', i.e. all is honey; food is honey (Wakhle, 2002). In Islamism, honeybees were entitled *al-Nahl* in the holy Quran's chapter (Surah). In Christianity, there are a number of references available about the importance of honey and bees (Webb, 2012). Similarly in Buddhism and Judaism, honey makes its symbolic significance (Nayik *et al.*, 2014).

The exact date of origins of honey for the man always remains a mystery. But the honey's use by man precedes about 40,000–8000 years ago, evident from rock art related to honey and bees found in Spain, India, Australia and South Africa (Crittenden, 2011).

Why Should the Quality of Honey be Maintained?

Honey consumption as a diet and medicine has high nutritional and therapeutic value (Table 3.1). As a natural sweetest substance, it is processed and manufactured by honeybees by collecting the nectar from the surrounding flowers (Ahed and Khalid, 2017). Having high nutritional value, i.e. 330kcal/100g it is also known for its fast-absorbing carbohydrate activities after its intake (Conti *et al.*, 2007; Kowalczuk *et al.*, 2017). Honey properties, namely the antioxidant, antibacterial, anti-fungicidal, hepatoprotective and anti-inflammatory are widely known (Noori *et al.*, 2014; Meo *et al.*, 2017). The health-supporting qualities associated with honey are attributed to biotin, niacin, folic acid, thiamine, phytosterols, polyphenols, tocopherol (all metabolites) and from various enzymes along with different coenzymes.

Natural honey also retains tropical impacts, including memory-boosting, in addition to neuropharmacological activities, like anticonvulsant, anti-nociceptive, antidepressant and anxiolytic properties. The polyphenol components present in honey can snuff out reactive biological oxygen species and counteract the oxidative stress and restore the antioxidant defence system of the cell (Rahman *et al.*, 2014). Apitherapy is a branch of alternative medicine which describes the medicinal use of honey along

TABLE 3.1

Some Important Medicinal Curative Properties of Honey and Component Responsible

Properties of Honey	Component Responsible	References
Antioxidant	Phenolic acid and flavonoid	Ahmed *et al.* (2018)
Anti-inflammatory	Flavonoid (i.e. galangin)	Testa *et al.* (2019)
Antibacterial, antiviral and anti-parasitic activity	Hydrogen peroxide, low pH	Meo *et al.* (2017)
Anti-mutagenic and antitumour activity	Phenolic acids, flavonoid	Ahmed *et al.* (2018)
Skin grafting and wound healing	Hydrogen peroxide	Goharshenasan *et al.* (2016) and Maghsoudi and Moradi (2015)
Improves serum testosterone levels	Vitamin B	Meo *et al.* (2017)
Increase sperm quality	–	Meo *et al.* (2017) and Fakhrildin and Alsaadi (2014)
Enhance fertility	Nitric oxide	Kishore *et al.* (2011) and Meo *et al.* (2017)
Nootropical activities	Polyphenol, flavonoid	Rahman *et al.* (2014)

with other bee's products. It is now seeking attention and has become a distinct area of research as alternative medicine because of various familiar preventive or curative methods of honey uses, for many disorder treatments, often shown in folk medicine.

The world honey market is expected to grow 2.4 million tons in the coming year 2022 (Global Industry Analysts Inc., 2016), and this anticipated growth lies on the believes and awareness about honey's natural origin and it is not the cause of obesity as associated with the other sugar sources. Like any other growing market, honeys too have a number of concerning issues, most of them related to quality of honey.

Characteristic Feature of Quality Honey

Honey is a sweet gift of nature, having varied constituents, which are affected due to honey's botanical nature, geographical origin, nectar flow's intensity, the climatic conditions, manipulations by the beekeepers, the way of handling, procedure of packing, the storage condition and time duration etc. (Thrasyvoulou *et al.*, 2018).

Honey can be categorised based on its:

- Origin, i.e. blossom honey, honeydew honey, monofloral honey, multifloral honey.
- Way of its harvesting and processing, i.e. strained honey, pressed honey, comb honey, extracted honey, crystallised or granulated honey, creamed honey and chunk honey.
- By its intentional use, i.e. table honey, industrial or baker's honey.

Criteria for quality of honey are mentioned in European Directive and the Codex Alimentarius Standard. The definition of honey according to European Directives states that honey is the substance which is produced by honeybee species, namely *Apis mellifera* (Thrasyvoulou *et al.*, 2018). However, Codex Alimentarius Standard states honey as nature's sweet, formed by honey bees using the nectar collected from flowers or from secretion and excretions of any other alive plant's part and of sap-sucking insects, respectively, which are accumulated, transformed by them through amalgamating it with certain own ingredients, dehydrate, store and then leave it in combs of hives to get ready for consumption after-ripening (Codex Alimentarius, 2001).

A number of different methods which are significant to assess the quality parameters present in honey have been given in the European Directive and in the Codex Alimentarius standard. Besides these, some other quality parameters based on modern methods, such as analysis of individual sugar, electrical conductivity and invertase number, have been increasingly used, mostly in the countries which are more industrialised; therefore, a more accurate characterisation for the given honey can be attained.

Honey is a supersaturated sweet, which is composed primarily of D-fructose, D-glucose, maltose and sucrose along with a variety of higher sugars. Besides these, a wide varieties of some minor components are also present in honey-like alkaloids, flavonoids/isoflavones, glycosides, phenolics, peptides/proteins, certain enzymes (invertase, amylase and glucose oxidase), carotenoid-like pigments, organic acidic components, products of Maillard reaction, minerals and vitamins (Pavlova *et al.*, 2018).

The physicochemical properties associated with honey are significant for determining the quality and origin of honey. The physicochemical characteristics of honey depend on the flora explored by the honeybees, regions of visit, beekeeping practices and environmental climatic variations. The honeys collected from varied sources have been analysed extensively for their physicochemical characters by various researchers. The main criteria for quality assessment include pH value, moisture, reducing sugars, sucrose presence, electrical conductivity, free acidity, ash content, diastase activity and hydroxymethylfurfural (HMF) content.

Quality Parameters and Their Compositional Criteria for Honey

The important quality parameters to assess the honey comprise water-insoluble content, moisture content, sucrose, sugar content (fructose and glucose ratio), free acid, electrical conductivity, diastase activity and HMF content. Other quality parameters are proline, ash content, specific rotation, invertase activity etc. (Table 3.2) (Bogdanov *et al.*, 1999).

Moisture

Water is among the main components assessing the quality parameters related to honey. The moisture content may vary from 15 to 23% in various honeys. Normally, honey extracted from fully closed combs contained the moisture below 18%. According to the Codex and European Directive, water content present in honey should be less than 20% with the exception of (*Calluna vulgaris* honey) heather honey that is allowed up to 23%. The moisture content of honey is related to climatic conditions, variety of the bees, the bee colony strength, the humidity and the temperature of air in the hive, the condition of processing and storage and the honey's botanical origin. It also affects some physical

TABLE 3.2

Compositional Criteria of Honey

Composition Criteria	Directive 2001/110 EU			Revised CODEX 2001
	Blossom Honey		Honeydew Honey[a]	
	General	Exceptions	General	General
Moisture percentage	Less than 20	Calluna and baker's honey less than 23; Baker's honey from Calluna less than 25	Less than 20	Alike no indication for baker's honey
Fructose + glucose percentage	More than 60	–	More than 45	Alike
Sucrose percentage	Less than 5	Medicago, Robinia, Hedysarum, Banksia, *Eucryphia* spp., Eucalyptus, *Eucryphia* spp. and Citrus more than 10 Lavandula and Borago less than 15	Less than 5	Alike
Water-insoluble percentage	Less than 0.1	Pressed honey less than 0.5	Less than 0.1	Alike
Electrical conductivity (mS.cm^{-1})	Less than 0.8	Chestnut, Erica, Arbutus, Eucalyptus, Calluna, Tilia, Melaleuca and Manuka	More than 0.8	Alike
Free acid (meq/kg)	Less than 50	Baker's honey less than 80	Less than 50	Alike
Diastase activity (DN[b])	More than 8	Baker's honey and honey with low natural enzyme content: more than 3 when HMF is less than 15 mg/kg	More than 8	Honeys with less natural enzyme content: > 3 DN
HMF (mg/kg[b])	Less than 40	Baker's honey Honeys of tropical climate and blends of these honey less than 80	Less than 40	Honeys of tropical climate and blends: <80

Source: Thrasyvoulou *et al.* (2018).

[a] Blends of honeydew honey with floral honey and honeydew honey.

[b] Observed after blending and processing.

features present in the honey, i.e. flavour, viscosity, maturity, crystallisation, specific weight and specific gravity (do Nascimento *et al.*, 2015; Pavlova *et al.*, 2018). It is an important factor that shows influence on the honey's lifespan. Honey having higher moisture content may be susceptible to the fermentation because of low osmotic pressure of sugars (not much powerful) to inhibit the growth of osmophile yeast which are present in high sugar concentration (Bogdanov and Martin, 2002). Therefore, chances of fermentation increase, with increase in water content (Jovanovic, 2015).

However, some researchers took into account the water activity that is responsible for the growth of microorganism. The water activity (aw) shows the quantity of free water present for the growth of microorganisms. Honey's water activity varies from 0.49 to 0.65; it can extend up to 0.75 for some honey types (Cavia *et al.*, 2004; Costa *et al.*, 2013). The water activity requirement for growth of microorganism is about 0.70 for moulds, 0.80 for yeast and 0.90 for bacteria. Water activity (aw) values less than 0.60 prevent the growth of fermentative osmophilic yeasts (Bogdanov, 2011b).

Sugars Content

Honey includes sugars (chiefly fructose and glucose) and fructo-oligosaccharides in small amounts and water. The sugars are accountable for 95–99% dry matter of honey, and fructo-oligosaccharides about 4–5% (Ajibola, 2015). Generally, fructose and glucose in honey are (32–44%) and glucose (23–38%), respectively. However, the proportion can vary in different honey. There are set criteria for entire glucose and fructose present that is more than 60 g/100 g for nectar honey and for honeydew it should not below 45 g/100 g. The glucose and fructose ratio in honeydew and blossom honey blends should be more than 45%. But the blends of forest honey and blossom honey are permissible only if blossom honey dominates over honeydew honey and glucose and fructose are not below the level of 60% and electrical conductivity should not exceed 8 mS/cm. This honey blend is produced, when from blossoms bees gather nectar and from plants they collect honeydew, during a common blooming period belonging to one and the similar geographical area. The honeydew honey of spruce, fir and pine contains lower glucose and fructose ratio that can be considered as an exception. The honeydew honey contains higher fructose:glucose ratio as compared to nectar honeys (Jakubik *et al.*, 2020).

Sucrose Content

Sucrose is among one of the most common disaccharides present in honey. High sucrose present in the honey may be related to honey adulteration, although this may be due to immaturity of honey, botanical origin and high nectar flux. Sucrose content should not be above the level of 5 g/100g with exception of some specific honey (European Council Directive 2001/110/EC). But there are some exceptions, i.e. Eucalyptus honey contained more than 4.2% and dandelion honey (*Taraxacum officinale*) sometimes contains sucrose content exceeding 5%. Sugars present in the honey are produced by the bee's transformation of sucrose present in the nectar by the activity of invertase enzymes from the salivary glands of the bees (Machado De-Melo *et al.*, 2017).

Electrical Conductivity

Electrical conductivity is the fundamental property of a substance to make a way for electric current. This property is straightway linked with the minerals, inorganic ions, and in some way due to proteins' presence, organic acids and other compo-nents which can behave electrolyte like sugars, polyols and pollen grains (Machado De-Melo *et al.*, 2017). Electric conductivity of honey is a significant parameter related to its botanical source and helps to determine the regular honey control as compared to the ash content. Generally, electric conductivity increases with an increase in the acid and ash content of honey, the higher the acid and ash component, the greater will be electrical conductivity. The standard limit of the honey sample not exceeds 0.8 mS/cm for the blossom honey and should not be lower than 0.8 mS/cm for the honeydew honey (Codex Alimentarius, 2001).

Acids

Honey contains various organic acids that are found associated with their corresponding gluconolactone in various equilibriums. They contribute about not beyond 0.5% of all solids in honey; still they are responsible for aroma, colour, taste, honey's preservation and for reducing the growth of microorganisms. The gluconic acid which represents 70–90% of all organic acids present is produced from glucose by the action of glucose oxidase present in the honey. Other organic acids seen in honey are as citric, acetic, formic, lactic, maleic, butyric, oxalic, fumaric, malic, pyroglutamic, pyruvic, succinic and tartaric acids (Mato *et al.*, 2003; Ahmed *et al.*, 2018) Out of these some are intermediate in the Kreb cycle and other enzymatic pathways (Machado De-Melo *et al.*, 2017). Generally, honey having acidity is darker in colour and storage of honey for longer increases acidity. Some honeys are characterised by low concentration of organic acid, therefore lighter in colour, i.e. Acacia, chestnut and meadow (Pavlova *et al.*, 2018). Acidity of honey is generally less than 50% for both blossom and honeydew honey (Thrasyvoulou *et al.*, 2018) with the exceptions of baker's honey. Some organic acids may act as the marker for distinguishing particular honey samples. The concentration of citric acid is used as a definitive quality parameter to distinguish two important categories of honey: floral and honeydew honey. Generally, honeydew honey contains higher concentration of citric acid, i.e. oak and pine honeydew and thymus honey (del Nozal *et al.*, 1998; Haroun *et al.*, 2012).

Diastase Activity

Diastase (amylase) destroys starch while added by the bees during the ripening process. It is most stable among other enzymes and resistant to heat; hence it is widely used as an important parameter of honey freshness and the value of diastase activity being regulated within several legislations, i.e. minimum value 8. It is defined as that the amount of α-amylase, which will convert 0.01 gram of starch to the prescribed endpoint in 1 hour at 40°C, is known as one unit of diastase activity. The results are indicated in Schade units per gram of honey and known as diastase number (DN) (Kuc *et al.*, 2017).

Hydroxymethylfurfural

HMF is an organic compound, formed as a breakdown product of simple sugars (such as fructose), which is formed in a slow and natural process in the course of honey storage, and much more rapidly while honey is heated (Machado De-Melo *et al.*, 2017). HMF can also be formed from the hydrolysis of oligo- and polysaccharides which can give hexoses. It can also form during the Maillard reaction. There are a number of factors that show impact on HMF formation, i.e. heating period, temperature, state how it is stored, pH and honey type. High content of HMF, i.e.

more than 100 mg/kg can be an indicator of honey adulteration (Batinic and Palinic, 2014). Therefore, the Codex Alimentarius Standard commission permitted upper limit for HMF is 40 mg/kg (whereas upper limit is 80 mg/kg for tropical regions honeys) to make sure that the honey is safe for consumption and has not heated extensively during its processing. The HMF formation rate is related to fructose:glucose ratio and it has been found that fructose possesses reactivity fivefold higher than glucose at pH 4.6, and a higher fructose:glucose ratio will speed up the HMF formation (Shapla *et al.*, 2018).

Water-insoluble Contents

Water-insoluble contents in honey are impurities present in honey, i.e. (wax remains, bee and larvae corpses, wood particles, pollen etc.), or minerals (dust, earth etc.). The amount of impurities comes in honey through improper honey extraction, filtering and processing, handling and storage techniques. The presence of impurities in honey makes it low quality in context of organoleptically and presentation. It also affects the preservation capacity and irregular crystallisation. In pure honey, impurities must be less than 0.1%. Honey is always present with some impurities. So there are probable chances of adulteration increases (Pourtallier and Taliercio, 1972).

Processing of Honey

Honey in the comb is the best condition for the honey storage as this is processed by bees themselves, whereas, outside the natural condition, it is sensitive to environmental conditions since it can lose its colour, flavour, taste and quality. There are several factors, i.e. microorganism, acidity, heating or sunlight which ultimately lead to deterioration of honey quality. The major threat to unprocessed honey (crystallised or liquid) is fermentation, caused by osmophilic yeast under the favourable condition of temperature (<30°C) and humidity (<50%). Furthermore, Honey obtained from combs of various apiaries may contain pollens, beeswax, dirt, dust and other unwanted materials that can encourage the fermentation, besides providing turbidity to honey. Therefore, these undesirable matters must be removed from the honey to sustain the quality and to enhance the shelf life of the product number of steps involved in honey processing like, i.e. uncapping, extraction, preheating, filtration, heating and thermal processing, cooling and packaging. Among these most important are filtration and thermal process or heating.

Simple Straining

The simple straining isolates the large wax particles and various suspended solid particles present. Usually, the method and the equipment vary according to the scale of the taken operation. In low size operations, it is normally executed with bags or cloth of nylon. However, in bigger size operations, the straining is done in combination with the preheating (till 40°C) in an encased tank having a stirrer (Wakhle and Phadke, 1995).

Filtration

In pressure filtration, the honey is filtered through an 80-μm polypropylene microfilter, between 50 and 55°C, so that beeswax liquefying can be prevented. But in large-scale processors, the filtration process expanded to various steps which include coarse filtration, followed by centrifugal clarification and then fine filtration and blending are done.

Thermal Processing or Heating

Honey processing is necessary to destroy sugar tolerant yeast present in honey which is responsible for honey fermentation, by reducing moisture content which prevents honey from granulation. However, it should be done in such a manner that it should not alter the other quality parameter of honey that is used for honey authentication such as HMF, diastase activities etc. The initial concentration of HMF varies in different honey samples, which are subjected to the climate behaviour of honey origin, apart from other factors (Subramanium *et al.*, 2007). Normally high viscosity and crystallisation of honey are two properties that can create difficulty in managing and processing of honey. Viscosity of honey is determined by many elements which include moisture, composition and temperature. Generally, it decreases with increase in temperature; a number of researchers found out the correlation between the glass transition temperature (T_g) and critical viscosity, which shows that the critical viscosity can be obtained at 10–20°C more than temperature of the glass transition temperature (Bhandari *et al.*, 1997). In a recent analysis, it is shown that the viscosity of honey remained unchanged upon gamma radiation treatment. Hence radiation treatment can be useful for microbial safety of the honey (Saxena *et al.*, 2014).

Crystallisation of honey can be prevented by (Bhandari *et al.*, 1999):

1. Storage of honey at −40°C (freezing temperature).
2. Heat treatment of honey to dissolve crystals and nuclei of crystals.
3. Filtration to remove air bubbles, dust and pollens.
4. Filling which is done at > 45°C to prevent incorporation of air bubbles.
5. Adding inhibitors, i.e. sorbic acid and isobutyric acid.
6. Regulating the ratio of glucose to fructose.

Conventional Methods

Conventional methods recommend preheating to 40°C, after that filtration of honey is done which is followed by indirect heating at 60–65°C, for duration of 25–30 minutes in a tube-shaped heat exchanger. Thereafter swift cooling is done immediately for preservation of flavour, colour, enzyme present and other organic matters (Wakhle and Phadke, 1995). Treatment of honey at 63, 65 and 68°C for the duration 35, 25 and 7.5 minutes, respectively, can be effective to kill the yeast microorganisms (Wakhle *et al.*, 1996). In high temperature, i.e. 80°C, short-time heating for duration of 60 and 30 seconds in transient and in isothermal phases, respectively, effectively vanishes the microorganisms which spoil honey grade and can be considered as gentle or appropriate treatment process (Tosi *et al.*, 2004).

Heating treatment at 140°C, for duration of 15 and 30 seconds in transient and in isothermal phases, respectively, found is very critical to cause a permissible quality loss (Tosi *et al.*, 2004). The heating of honey at 55–75°C for 15–25 minutes and storing the honey at ambient room temperature (20–23°C) for 26 weeks showed storage time effect but no heating treatment effect on diastase number and invertase, whereas heating and storage duration shows significant influence on the HMF content (Hasan, 2013). Besides this, there is a different thermal requirement of different honeys that can have less impact on other quality parameters of honey.

According to Blidi *et al.* (2017), there is significant alteration of both characteristic specifications, i.e. HMF and diastase activity for honeys of different botanical origins, under heating at 65°C for 6 hours. However, heating at 45°C for 24 hours was found to be the least severe treatment with regards to the variation of HMF and diastase activity. Furthermore, they found pine honey as the most resistant sample which shows HMF formation and multifloral honeys as the least altered enzymatic activity through the whole thermal process observed.

Modern Methods

Keeping in view the rising demand of honey due to its numerous health benefits and increasing interest of people in natural products, the quality of honey becomes very important to sustain the trust of the people in this natural product. Conventional methods significantly affect the different characteristic specifications of honey like HMF content, invertase and diastase activity, and sensory characteristics of honey. Therefore, alternative methods are gaining popularity over conventional thermal processing. These include microwaves heat processing, ultrasound processing for liquefying, membrane processing and infrared heat processing (Subramanian *et al.*, 2007).

*   **Microwave heat processing:** Microwaves are the waves that belong to a wide group of electromagnetic waves having 300 MHz to 300 GHz varied frequency range. Unlike conventional heating, microwaves go deep in the substance, interreact with it, which create heat that cause heating of substance rapidly. Polarity of molecules, as in water, leads to its heating quickly as when subjected to microwaves, because of molecular friction owing to the dipolar rotation, which occurs in an alternating electric field. Microwave processed honey heated at 70°C can be stored safely at room temperature, which provides insufficient count of yeast to initiate fermentation. This shows significant resistance towards spoilage and granulation as compared to the unprocessed raw honey. However, there may be the colour change and loss of diastase activity about 37.5% which further be prevented by storage of honey at 4°C (Ghazali *et al.*, 1994). There is a different combination of heating periods and micropower magnitude to attain the commercialised yeast reduction activity of honey. But heating duration of 15 seconds at high power magnitude (i.e. 16 W/g)is helpful in maintaining the lower level of HMF value (i.e. 3.8 mg/kg) and greater diastase activity (i.e. 12.0)

(Hebbar *et al.*, 2003). Dimiņs *et al.* (2019) reported that HMF level rises and invertase activity rapidly declines, while using 450 Watts microwave power for 10 s. The decrease in invertase activity is somewhat higher with greater microwave power. However, there is greater increase in HMF content at small microwave power in contrast to invertase activity decreases.

- **Infrared heat processing of honey:** The application of infrared in process-ing of honey was conducted by Hebbar *et al.* (2003). The studies showed that processing honey continuously for 2, 3, 4, 5 and 8 minutes causes substantial reduction in yeast count. However, there is 220% increase in HMF content and 37% decrease in enzyme activity, when processing done for 5 minutes results in 85°C product temperature. Duration of 3–4 minutes is found most suitable to attain a commercialised product that possesses all the regulatory characteristic preconditions specification, i.e. HMF content (≤40 mg/kg), diastase activity (DN ≥ 8), yeast and moisture level. There is further need to study the impact of infrared processing of honey belonging to different botanical origins on its quality criteria due to lack of scientific literature.

- **Ultrasound processing of crystallised honey:** Normally, raw honey is processed to obtain the desired quality, which is mostly in non-crystallised form. However, for raw crystallised honey, it can be subjected to ultrasonic treatment. Ultrasonic waves include high-frequency waves in comparison to the maximum limit of human ear sound wave hearing capacity. They can penetrate through water without losing intensity, which leads to ther-mal and mechanical transitions in the matter that undergoes exposure and consequently also leads to certain alterations in unicellular organ-isms. Consequences of ultrasonic waves on quality of honey conducted by Thrasyvoulou *et al.* (1994) on granulated samples of honey (both for blos-som and honeydew honey) subjected to the treatment at 23 kHz ultrasonic waves and complete liquefaction achieved within 18–25 minutes ultrasound exposure processing. This liquefaction energy requirement ranged from 0.1056 to 0.1466 kWh, whereas sample attains maximum temperature at 76–82°C. However, time requirement for liquefaction is subjected to initial crystallised state and character of honey. Moreover, rise in HMF content and decrease in diastase activity are comparatively lower in ultrasonic treat-ment and at this optimised condition of power, ultrasound processing (i.e. in volume of 60 mL, treatment time 8 minutes and amplitude of 60%) and controlled temperature condition (within range from 65 to 95) destroy yeast better than thermal processing and contained higher nutritional value in comparison to thermally processed honey samples (Janghu *et al.*, 2017).

- **Membrane processing of honey:** The membrane processing of honey is a thermal process which involves ultrafiltration of honey by using filtering membrane technology (Figure 3.1). Normally conventional processing may lead to loss of important quality characteristics of honey that is reduction in enzyme activity. Therefore, anticipated quality of honey can be main-tained by using filtering membrane technology. Benefits derived from mem-brane processing are to obtain ambiguity or crystallisation free honey, with lowered viscosity and commercially issueless product (Subramanian *et al.*, 2007). Ultrafiltered membrane having a molecular weight cutoff 10,000

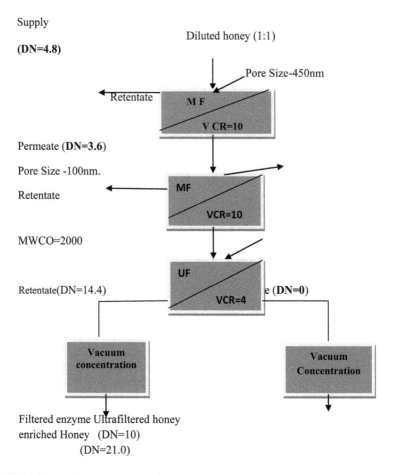

Supply

(DN=4.8)

Diluted honey (1:1)

Pore Size-450nm

Retentate   M F

V CR=10

Permeate **(DN=3.6)**

Pore Size -100nm.

Retentate ←   MF

VCR=10

MWCO=2000

UF

Retentate(DN=14.4)   e **(DN=0)**

VCR=4

Vacuum concentration

Vacuum Concentration

Filtered enzyme Ultrafiltered honey
enriched Honey (DN=10)
               (DN=21.0)

FIGURE 3.1   Membrane Processing of Honey. (Barhate *et al.*, 2003; Subramanian *et al.*, 2007: Adapted with the permission from www.tandfonline.com.)

perhaps eliminates the microbes completely, and making honey as micro-biologically unharmed additive product (Itoh *et al.*, 1999). But honey free from useful enzymes and protein can't be considered as suitable for con-sumption as food. The total permeation flux, in addition to the discharge rate of sugar, rises with operating temperature rise and by increasing honey dilution at continuous supply flow velocity and pressure applied. Moreover, it is also enhanced as the molecular weight cutoffs of the used membrane increased (10,000, 30,000 and 150,000). According to Barhate *et al.* (2003), the effectiveness of ultrafiltration (UF) membranes in enzyme rejection and yeast cell elimination is reported for 50% diluted honey with water with different MWCO. It is also observed that permeates flux through 25,000, 50,000 and 100,000 MWCO UF membranes completely reject amylases and lose diastase activity completely. The average discharge rate achieved with used membranes lies within the range of 0.90 and 1.15 kg/(m²h) (Figure 3.2).

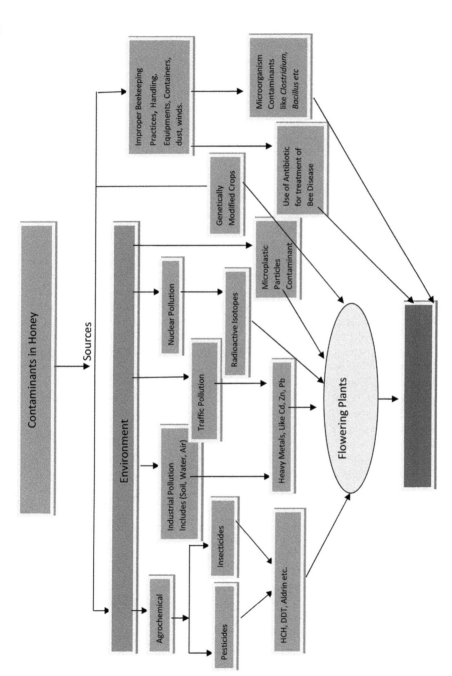

FIGURE 3.2 Sources responsible for honey contaminations.

Furthermore, clarified honey and various enzyme-supplemented honeys can be produced by combining microfiltration and UF membranes while processing. The MF membrane use (100 nm aperture size) in this procedure will eliminate the yeast cell along with other microbes completely. Because of high discharge flux through the membrane, the nutritional factors are left in the product stream (permeate). After that, using 20,000 MWCO, UF membrane will separate the microfiltered honey into two parts termed regular ultrafiltered honey and enzyme-enriched honey. The constraint of membrane processed honey is removing water content which is added before membrane processing for dilution; therefore, vacuum concentration done at lower temperatures is important for reduction of the heating harm to the honey (Subramanian *et al.*, 2007).

Honey Adulteration

Adulteration of honey is a matter of great concern to the people in respect of food safety, health benefits derived from the honey, fraudulent protection, maintaining the consumer trust and moreover to protect from hazardous effects of cheap added material in the honey. Authenticity of honey is with respect to botanical or geographical origin, i.e. mislabelling of honey origin and adulteration with sugars and syrup (as adulterant). According to the Food Safety and Standards Act of India (FSSAI), definitions of adulterant are any substances that are used in food items to make it harmful or substandard or adding irrelevant material. In honey, it can be of two types, on the basis of way of manipulation, i.e. direct or indirect adulteration. Direct adulteration can be done by straight inclusion of sucrose syrups that are obtained from sugar beet, corn syrup that is rich in fructose, maltose syrup or by addition of industrial sugar, i.e. fructose and glucose, starch-derived syrups through heating and treatment of honey with acid or enzyme, whereas indirect adulteration done by excessive noshing of the bees present in the colonies by syrups in the course of prime nectar flow period (Tura and Seboka, 2020).

The main reasons for adulteration are as follows:

- Intended to earn higher cash income by making large volumes.
- Environmental pollution increase (Zabrodska and Vorlova, 2014).
- Spread of various diseases globally in honeybee has led to decline their populations (Zabrodska and Vorlova, 2014).
- People's interest in natural products increases the demand of honey extensively and makes it a scarce commodity.

Honey adulteration, besides reducing the quality of food items, also results in various health-related issues. Therefore, testing for quality assessment of the honey is an essential requirement to assure consumer health along with their protection against fraudulent activities.

Honey Adulterants

Adulterants used in honey are usually low-cost sugars and commercial syrups. The most common well-known adulterants belong to sugarcane and sugar beets that include various syrups like corn syrup, sucrose syrup, glucose syrup, inulin syrup

TABLE 3.3

Some Commonly used Honey Adulterants

Adulterants	Adulteration Method: Direct/Indirect	Countries	References
Cane sugar	Both direct and indirect	–	Li *et al.* (2017)
Corn syrup	Direct	–	Olivares-Pérez *et al.* (2012)
Palm sugars (Jaggery syrup)	Direct	India	Fakhlaei *et al.* (2020)
Invert sugar	Direct	–	Veana *et al.* (2018)
Rice syrup	Direct	China	European Chemical Agency (2010)
Inulin syrup	Direct	–	Fakhlaei *et al.* (2020)
Blending	Mixing of pure and cheap honey	China, Venezuela	Cordella *et al.* (2005) Irawati *et al.* (2017)

with high fructose, HFCS and inverted syrup etc. Honey adulterated with different sugars results in transformation of the chemical as well as their biochemical characteristics of honey, i.e. electrical conductivity, enzymatic activity, and contents of particular mixtures (Table 3.3) (Fakhlaei *et al.*, 2020).

Methods for Honey Authentication or Adulterations

Assessing the honey quality and its authenticity is a somewhat complex task, a wide variety of methods are employed to detect the authenticity of honey. However, all the available methods rely primarily on physicochemical characteristics of honey. Therefore, to maintain the quality of honey, different scientific techniques are applied to measure originality of honey and researchers are still exploring or modifying the existing methods to detect these honey frauds and restore the consumer's interest. The method of honey authentication can be classified into two broad categories:

- **The conventional methods** include physicochemical parameters and melissopalynology.
- **Modern methods** include mass spectrometry, chromatography, nuclear magnetic resonance, infrared spectroscopy and molecular techniques (Chin and Sowndhararajan, 2020).

Conventional Methods

Physicochemical quality parameters, i.e. pH, free acidity, proline, diastase activity, enzymatic activity, ash content, sugar content, HMF content and moisture content, can give important details about honey's origin and its quality. The standard methods used for detection of these parameters are available after harmonised methods (International Honey Commission, 2009).

- Abbe refractometer is used for measurement of moisture content. Electronic refractometers are utilised for measurement. The moisture amount is usually below or 19% to secure a greater lifespan for honey that will be according to the majority of the commercial honey samples. The RSDR level is given from 0.8 to 2% throughout the total determination scale.
- The electrical conductivity of a sample is measured by taking 20 g honey dry material and dissolve it in 100 ml of distilled water with an electrical conductivity cell. Reading depends upon the amount of ash and content of acid present in the honey, i.e. if their content will be higher, the conductivity will be higher.
- The pH of different samples of honey is measured with a pH meter and the solution is titrated further with a solution of 0.1 M sodium hydroxide (NaOH).
- HMF content is measured with the help of reverse phase HPLC coupled with ultraviolet detector by taking filtered, clean and clear, dissolvable honey solution. The signal obtained is compared with those from standard solution (known concentration).
- The diastase activity is analysed in diastase number (DN) by using Schade units. It can be illustrated as: It is the enzyme's activity calculated in1 g honey sample that hydrolyses starch of 0.01 g at 40°C within 1 hour is known as one diastase unit. The diastase activity is calculated with the help of Phadebas test (Siegenthaler, 1975; Bogdanov, 1984) and methods followed after and Schade *et al.* (1958).
- Sugar content is calculated with modified method of the Lane and Eynon (1923), which involves Soxhlet's modified Fehling's solution reduction by using titration against a honey solution at boiling point, methylene blue can be used as an internal indicator. To calculate the evident amount of sucrose present, variation in mass of invert sugar after and before is multiplied by 0.95. The sugar content can also be determined by HPLC Bogdanov and Baumann methods (1988) or by GC (gas chromatography).
- Invertase activity is represented in units, i.e. one unit is illustrated as the micromoles of substratum spoiled per minute and represented as per kilogram of honey. This can also be represented in invertase number. Invertase activity of honey samples can be determined by using U. Siegenthaler (1977) methods.
- Proline reacts with ninhydrin and forms a complicated chroma and then 2-propanol is added, the disappearance of taken sample solution and a proline standard solution at a higher wavelength is calculated and the ratio is obtained, which is used to find the proline level (Ough, 1969).

Melissopalynology is a classical technique used for determining geographical and floral origin of honey. This technique is used extensively to verify these claims and this method is in routine practice as a part of food standard assurance procedure especially in the European Union (European Council, 2001). In melissopalynology, pollen grains seen in blossom honey are identified with the help of microscopic analysis (Kale Sniderman *et al.*, 2018). Sometimes the melissopalynology not be suitable for certain varieties of honey as of citrus, it is due to variable pollen content and the little pollen count. Limitation that may occur for the honey belonging to the same floral source from the different geographical origins shows little variation in some specific pollen content (Kale Sniderman *et al.*, 2018). Besides this, in some instances, the pollen grains get filtered and found in bee's honey sac and sometimes are mixed with the intension to fraud in honey.

Modern Methods

Chromatographic Techniques

These techniques are in wide use for identification of adulterated honey. A number of workers used different chromatographic methods to calculate sugar content, different amino acids and also for phenolic and flavonoid components (Perna *et al.*, 2013; Karabagias *et al.*, 2014). Gas chromatography (GC) is applied to calculate different saccharides (mono-, di- and tri-) present in honey using relatively high resolution and sensitivity and it also helps to expose the presence of different adulterants. GC can easily be detected adulterants like maltose, isomaltose and HFCS along with quantities and ratios. The presence of HFCS can be found by detecting DFAs (difructose anhydride) formed during heating of sugars and result in production of lower volatile components and larger fraction of non-volatile materials, i.e. 90–95% which also includes DFAs (Saito and Tomita, 2000). Thin-layer chromatography (TLC) can be used to calculate the purity in the given material and is carried out by using a plastic or glass sheet or aluminium foils having adsorbent material thin layer on them (stationary phase). The stationary phase is normally cellulose, aluminium oxide (alumina) or silica gel. When sample is put on the plate through capillary motion, the mobile-phase solvent or solvent mixture is drawn up on the plate. As different components of analytes have a different rate of rising, separation is attained on TLC plate. After performing this technique, patches on plate are figured out and analysed to calculate the results, i.e. the length of movement of different constituents of sample is divided by the complete length of mobile phase (length of mobile phase should not be equal to length of stationary phase), the ratio obtained is retardation factor (Rf). The HPTLC method used is based upon the sucrose level and fructose/glucose ratio to find adulterants present in taken honey (Table 3.4) (Puscas *et al.*, 2013).

Isotope Ratio Using Mass Spectroscopy

The stable isotope ratio is applied to check the honey adulterants. Honey-producing plants and sugar beets come under C3 category plants whereas plants of corn, sugar cane, along with other adulterating syrups plants come under C4 category. The difference in metabolism which produces the ^{13}C isotope is due to difference in their photosynthetic pathways of these plants. $^{13}CO_2$ is slow reacting and depleted substantially in C3 plants as compared to C4 plants in the course of CO_2 fixation. Therefore, it is feasible to detect the adulteration caused scaling up the amount of C4 plants sugar, which varies the ^{13}C level, i.e. honey of C3 category plants is -22 to -33delta0/∞, -10 to -20delta0/∞ for C4 category plants honey and -11 to -13.5delta0/∞ in honey of CAM (Crassulacean acid metabolism) plants like cactus and pineapple. By adding sugars from C4 plants, the delta^{13}C value of honey is altered. The honey having

TABLE 3.4

Chromatographic-Based Technique used in Honey Adulteration or Authentication

Chromatographic-Based Method	Component	Adulterants	Comments	References
Gas chromatography	Maltose/isomaltose ratio	High-fructose corn syrup	Ratio higher than 0.51 indicates adulteration	Doner *et al.* (1979)
Chromatography	Sucrose level	Commercial sweetener	Sucrose level slightly higher in adulterated	Prodolliet and Hischenhuber (1998)
High-performance anion-exchange chromatography (HPAEC-PAD)	Sugar components	Bee feeding sugar syrup	Modification in honey quality	Cordella *et al.* (2005)
HPLC-diode array detection (DAD)	2-Acetylfuran-3-glucopyranoside. Adulterant component in rice syrup	Rice syrup	2-Acetylfuran-3-glucopyranoside. Adulterant component in rice syrup	Xue *et al.* (2013)
LC with a pulsed amperometric detector (PAD); GC with flame ionisation detector (FID)	Glucose and fructose; di- and trisaccharides	–	For authenticating unknown commercial honey and appearance of trisaccharides in fir honey	Cotte *et al.* (2004)
Liquid chromatography–electrochemical detection (LC-ECD)	Amino acid	Mixing of acacia honey with rape honey	Chlorogenic acid contents were high in acacia honey; ellagic acid low in rape honey	Iglesias *et al.* (2006)
HPTLC	Fructose glucose ratio and sucrose content	Sweeteners	Altered composition	Puscas *et al.* (2013)

delta^{13}C value less than delta 23.5‰ can be considered as adulterated honey. The honey extracted protein may be taken as an internal reference to detect the presence of adulterants in sample honey, the similar delta^{13}C value of extracted protein will remain same. The variation level accepted for honey and its relative extracted protein in delta^{13}C value is −1delta0/∞ of minimum variation, which gives the international standard value as 7% of added C4 plants syrups (Padovan *et al.*, 2003).

NIR Spectroscopy

It is an important method to find the honey adulteration of samples and because of its rapid and non-destructiveness make it suitable diagnosing method used in checking the honey's quality. This system is used to procure NIR spectrum between the wavelengths of 400–2500nm under reflectance mode. NIR spectrometer is used to obtain sample to sample spectra, composition to composition detection and to measure the content of adulterants in the samples of honey (Sivakesava and Irudayaraj, 2000; Kumaravelu and Gopal, 2015; Tura and Seboka, 2020). Riswahyuli *et al.* (2020) used ATR-FTIR (attenuated total reflectance-Fourier transform infrared) spectroscopy in combination with the chemometrics (chemical technique which uses maths) that enable us in classification, in regression model development and in generation of grouping which helps to differentiate between unauthenticated and authentic Indonesian wild honey.

Nuclear Magnetic Resonance Spectroscopy

It gives significant details about the structure for different types of food constituents, which make it one of the important scientific techniques to detect honey genuineness since it gives the main focus to both chemical and structural characterisation. The common experiments use ^1H or ^{13}C spectrum of nuclear magnetic resonance (NMR) which is one dimensional and the classical $^1H^{13}C$ spectrum HMBC (heteronuclear multiple-bond correlation) is two dimensional. NMR spectroscopy helpful in identification of individual compounds and confirmed the important markers for honeys of different botanical origin (Chin and Sowndhararajan, 2020). It is also applied to check honey adulteration caused by cheap sugars addition. Bertelli *et al.* (2010) identified the honey adulteration caused by deliberated addition of various commercial sugar syrups in different concentrations by taking their NMR (1D and 2D) along with statistical analysis of various observations. The use of low field ^1H NMR is also found suitable to categorise eight different Brazilian honeys according to their botanical origin and to identify honey adulterated with different concentrations of HFCS which is 0% for pure honey and it can vary upto 100% for pure high-fructose corn syrup (Ribeiro and co-workers, 2014a, b).

Molecular Techniques

Honey is known to contain carbohydrates largely and little of protein (0.2%) which mainly comes from plant's nectar and bees (White, 1978; Lee *et al.*, 1985). Honey proteins are present in the enzymes form, primarily, including invertase, diastase, glucose oxidase and amylase. Besides this, acid phosphatase and catalase may also be seen according to their floral source and in recent times, proteolytic enzymes are also detected in honey. Various proteins present in honey have varying molecular weights which depend upon the species of bees to which honey belongs. Thus, different molecular techniques like real-time PCR and SDS–PAGE were applied to recognise and to check genuineness of the honey. It is used to distinguish honey originating from various types of bees, to confirm that protein constitutes are of bee origin and distinguish honey originating from native bees and honey originating from foreign bees (Chin and Sowndhararajan, 2020).

Honey Contaminants and their Potential Health Effects

Human beings have been gathering and using honey for food, for health benefit and, due to increasing consciousness for natural products, made it popular in existing and emerging markets and therefore health-conscious consumers and increase

TABLE 3.5

Honey and Beekeeping – Sources of Contamination

Types of Honey Contamination	Contaminants
Environmental contaminants	Heavy metals like cadmium, mercury and lead
	Radioactive isotope of various elements
	Polychlorinated biphenyls and other organic pollutants
	Pesticide like bactericides, herbicides, insecticides and fungicides
	Disease-causing bacteria
	Genetically modified organisms
Beekeeping contaminants	Fat-dissolving synthetic compounds as acaricides and non-toxic compounds like residuals of essential oil and organic acids
	Bee brood disease's antibiotics like chloramphenicol, sulphonamides, tetracyclines and streptomycin
	Chemical used on wax moths like paradichlorobenzene and chemical repellents

Source: Al-Waili *et al.* (2012).

in globalisation is carrying on the honey markets and its production. However, this honey market is also prone to contamination originating from a wide variety of sources in environmental. There are various factors which affect the quality of honey and may cause potential health hazards. These may include from beehive technology to increase in environmental pollution. Honey along with other bee products are polluted via different contamination sources (Figure 3.2 and Table 3.3). Honey contaminants can be Environmental Contaminants or Beekeeping Contaminants. Environmental contaminants are bacteria, pesticides, radioactive materials and heavy metals (Al-Waili *et al.*, 2012) (Table 3.5).

Pesticides

The use of pesticide has increased manifold in recent years, to protect the crops from getting damaged due to pests (Gomez-Perez *et al.*, 2012). Most pesticides on large scale are used in pulverised form and get dispersed, which pollute the food, water, soil and air in the surrounding area (Tette *et al.*, 2016). Bees because of their foraging behaviour come in contact with many pollutants, including pesticides and bring these pollutants to hive. Pesticide residuals may be fungicides, acaricides, insecticides, organic acids, bactericides and herbicides and mostly their uses are prohibited due to their widely known menaces to the health of human-like carcinogenic effects. Alam *et al.* (2017) found 84 pesticides from 18 honey samples from the North Lebanon region. The ten most common pesticides found include Pendimethalin, Diflufenican, Pyraclostrobin, Diuron, Penconazole, Hexachlorobenzene, a-endosulfan etc. Therefore, besides being impact on human health, honey also acts as an important bioindicator of environmental pollutants, i.e. pesticides. Besides this, there are many toxic substances used to control the honeybees disease like varroatosis and be ascospheriosis and includes taufluvalinate, acaricides, flumethrin, celazole, coumaphos, amitraz and bromopropylate that causes risk of contamination of honey along with other hive products (Korta *et al.*, 2002). There are a number of pesticide reported to identified in honey samples from different regions of world like HCH, DDT, Lindane, Aldrin, *cis*-chlordane, *trans*-chlordane, oxy-chlordane etc. (Al-Rifai and Akeel, 1997; Choudhary and Sharma, 2008; Yavuz *et al.*, 2010). The health hazards of pesticides to humans are numerous and depend upon the chemical's toxicity, duration and magnitude of exposure (Lorenz, 2009). It can cause genetic changes, birth defects, mild skin irritation, tumours, endocrine disorder, nerve and blood ailment and more severely also leads to death or coma. Prolonged use of organic pollutants such as DDT, heptachlor, chlordane, Aldrin, endrin, dihedron etc. can impact endocrine, reproductive and immune system and chronic exposure can cause neurobehavioral disorder, cancer, mutagenic effects and infertility (Ritter *et al.*, 2007; Lim *et al.*, 2010).

Antibiotic Contamination

Antibiotic contamination of honey is due to their use in controlling the bacterial disease of honeybees. Antibiotic contaminants originate from inappropriate practices of beekeeping and also from the environment. There are reports of a number of antibiotic's presence in honey like oxytetracycline and chloramphenicol residues, erythromycin, lincomycin hydrochloride, doxycycline, chlortetracycline etc. (Ortelli *et al.*, 2004; Saridaki-Papakonstadinou *et al.*, 2006; Granja *et al.*, 2009; Adams *et al.*, 2009; Chen *et al.*, 2001). Kumar *et al.* (2020) detected higher limit of erythromycin and oxytetracycline above acceptable levels set by FSSAI in honey samples from India. The antibiotics used in beekeeping practices is unlawful in some countries of the EU and also there are no set MRLs (maximum residue limits) of antibiotics for honey as per European Community regulations, which means antibiotic adulterated honey is prohibited from market selling (Forsgren, 2010). While certain nations like the United Kingdom, Belgium and Switzerland have set limits for antibiotics' (the presence of antibiotics above a certain level in honey is considered noncompliant) presence in honey, the permissible range of which falls from 0.01 to 0.05 mg/kg for every group of antibiotics (Al-Waili *et al.*, 2012). Residues of antibiotic present in honey are becoming a major cause of concern for consumers. Certain specific drugs can cause direct toxicity in consumers while some others are capable of causing allergic or hypersensitivity reactions (Velicer *et al.*, 2004). These antibiotic residues cause many long-term effects which comprise microbiological hazard, reproductive effects, carcinogenicity and teratogenicity. It can also result in microbiological effects, i.e. consumption of antibiotics along with food can result in generation of antibiotic concentration gradients in humans to which bacteria are exposed and develop resistance to antibiotics.

Heavy Metals

These metals are present in the soil as well as in air, they come primarily from industrial and traffic pollution and act as impurity in the hives of bees and in their products. The heavy metal in honey can also come from improper storage or equipment-related contamination, handling or processing (Gonzalez Paramas *et al.*, 2000). Even the containers used for storage and the galvanised tools with Zn can also contaminate honey (Al-Waili *et al.*, 2012). The principal toxic heavy metals studied are cadmium (Cd) and lead (Pb) commonly. Lead present in atmosphere primarily comes from smoke of vehicle that can disperse in the air and after that nectar and honeydew directly.

While Cd is released from metal industry and incinerators, contaminate nectar and honeydew when transported from the soil and plants. Bartha *et al.* (2020) detected the presence of some heavy metals like copper (Cu), zinc (Zn), cadmium (Cd) and lead (Pb) in various multifloral honeys belonging to Romania's industrial area, which is among the highest polluted zones of Eastern Europe. The presence of heavy metals in food is increasing and it is the gateway for entry into living bodies and can cause various hazards to human health. The heavy metals cause a number of health problems like deregulation of the immune system (Kim *et al.*, 2017) and cause various serious diseases comprising neurological ailments, cancers and cardiovascular diseases (Mates *et al.*, 2010; Tchounwou *et al.*, 2012; Lamas *et al.*, 2016). Heavy metals usually accumulate and biomagnify in the trophic chain, and when the concentration reaches above the maximum permission limits it becomes toxic (Mejias and Garrido, 2007).

The Radioactive Isotopes

The commonly reported radioactive isotopes among various honeys are potassium (^{40}K) and cesium (^{137}Cs), the first isotope originates naturally and the latter was originated from the accident that occurred in 1986 at atomic power station in Ukrainian (Chernobyl). Becquerel (Bq) per kg is used to indicate radioactivity. The ^{40}K radioactivity ranges between 39and 123 Bq/kg in various honeys reported from Poland (Borawska *et al.*, 2000). ^{137}Cs having 30 years half-life is the studied commonly after the Chernobyl atomic power accident. There are various studies undertaken to detect the existence of ^{137}Cs, concentration in different honey samples. The upper value permitted is 370 Bq/kg for milk in the EU and 600 Bq/kg was set for all other products (Bogdanov, 2004). Handa *et al.* (1997) studied the radioactivity due to different isotopes detected in honeys originated from Japan, China, Hungary, Italy and Australia and reported the presence of 226Ra, 214Pb, 214Bi and 40K, 137Cs and 134 Cs. Borylo *et al.* (2018) reported the ^{210}Po activity concentrations in honey samples collected from Northern Poland. It is also evident from these studies that the honeydew honey has more concentration of polonium as compared to nectar honey. The main health concern for consumers because of long-term exposure is the development of cancer. Consuming food having radioactive isotopes can increase the radioactivity amount within a person and hence radiation exposures within them increase and potentially increase radiation exposure associated health risk.

Microorganism

The presence of microorganism in honey can also affect the quality of honey. The microorganisms present in honey and in the combs of hives include yeast, bacteria and moulds. The possible sources of the microbial contamination are winds, dust, pollens, intestine of honeybee, human, equipment and containers. The microorganism can't flourish in the honey, it is because of its viscous behaviour, low water level and as a result of its antimicrobial activity properties (Snowdon and Cliver, 1996; Snowdon, 1999). But there are reports of microbial presence in honeys that might be due to recent contamination. Usually, bacteria which come in the honey collected aseptically lose their potentiality within 8–24 days if storage is done at 20°C (Olaitan *et al.*, 2007). But certain microorganisms which form spores could survive the low temperature of stored honey. These are *Clostridium perfringens, Clostridium botulinum* and *Bacillus cereus* spores that survived at the temperature of 25°C in stored honey. Contamination of honey with spore of has been reported in many countries, i.e. Argentina, California, Japan, Finland (Nakano *et al.*, 1992; Schocken-Iturrino *et al.*, 1999; Nevas *et al.*, 2002; Al-Waili *et al.*, 2012). The presence of *C. botulinum* spores in honey can cause infant botulism as spores in the digestive tract of infants increase through multiplication, which generate the toxin (botulinum) as and when it is given to infants or newborns.

The Transgenic Crops

Plants that contain the desired alignment of genetic material produced by using the modern biotechnological techniques are transgenic crops. In some countries, genetically modified crops are grown, i.e. rape and maize and might be cause of concern for beekeepers and can harm the bees also (Williams, 2002a,b). GM crops are commonly grown and accepted in some countries like the United States and Canada; however, the consumption of GMO-contained food is unlawful in European Union (EU). According to the Court of Justice present in European Union (EU), the pollen present in honey urged to be considered in the category of food ingredient and hence it comes under the range of 1829/2003/EC Regulation (Żmijewska *et al.*, 2013). The honey should be totally free from the GM pollen which is unauthorised in the EU, and if it contains pollen from authorised GM crops it must be labelled above the range of 0.9% as compared to whole content of present pollen (Żmijewska *et al.*, 2013).

Microplastics

Microplastics are fine particles of plastic normally about 5 mm (0.2 in.), which are present in the surrounding due to increasing pollution of plastic. It occurs in various forms like products of cosmetics, synthetic clothes, bags of plastic, bottles etc. Microplastics can be primary microplastic and secondary microplastic. Primary microplastic includes microbeads present in products used in personal care, industrial use of nurdles for manufacturing and use of plastic fibres in synthetic textiles. It comes in the environment by using personal care products (being washed with water and wastewater systems from households), spills dispersal in the course of manufacturing, transport, or abrasion caused due to washing (e.g. laundering synthetic textiles clothes). Secondary microplastics are breakdown products of larger plastic because of weathering and through the wave action, abrasion through wind and ultraviolet (UV) radiation from sunlight. Therefore, there are potential chances of food product contamination with microplastics. There are few reports available for contamination of food products, including honey with microplastic particles (Liebezeit and Liebezeit, 2015; Diaz-Basantes *et al.*, 2020). However, there is a need to investigate this contamination.

Conclusion

Whether it is quality honey production, honey processing or different mechanisms of adulteration detection advancement, refining of used techniques and inclusion of all the best practices available with the time, should always be of great significance for perpetual development of this industry and also for the growth of manpower associated with it. The different parameters of honey are important for quality assessment of honey and a number of factors affect these characteristics like botanical origin, temperature, climatic condition, humidity, storage condition, honey processing, equipment or container used. Although there are set quality criteria for honey qualities, but these are not applicable to all honey types. Therefore, these factors must bear in mind while assessing the quality of honey and there is a need to evaluate and categorise the data regarding various quality parameters of different honey types that can make it easy to assess the quality. Similarly, processing of honey is an essential requirement for removing undesirable material and microorganisms. It should be done in such a way to achieve the desired goal of processing and without affecting quality parameters at the same time. Although a number of methods are available, but some of them still need further studies for their extensive use, while others are tedious.

There are a number of methods available for honey authentication and for detection of its adulteration. In addition to conventional analytical methods, there are a number of modern methods available. Normally, it is somewhat easier to find out botanical or geographical origin of honey by conventional methods as well as with more precise modern techniques like NMR and molecular techniques. But detection of adulteration in honey seems a somewhat complex task and there is a need to further improve the existing modern methods and develop some novel techniques to detect different adulterant types in honey. Furthermore, there is also need to study the effect of adulterants on quality of honey that can pave the way to explore more precise ways to detect honey. Similarly, the presence of contaminants in honey comes from both the environment and from beekeeping practices. Measures should be taken on government level and also by beekeepers, i.e. there should be stringent laws to reduce and manage the sources of contamination and farmers should also be encouraged to minimise the pesticides used on the crops and adopt organic farming. Moreover, in beekeeping practices, there should be no use of chemical alternatives such as biological controls should be taken into account for combating different bee's diseases and besides this, setting of the beehives in the pollution-free area should be encouraged.

REFERENCES

Adams, S. J., Fussell, R. J., Dickinson, M., Wilkins, S., and Sharman, M. 2009. Study of the Depletion of Lincomycin Residues in Honey Extracted from Treated Honeybee (*Apis mellifera L.*) Colonies and the Effect of the Shook Swarm Procedure. *Analytica Chimica Acta* 637(1–2): 315–320.

Ahed, A., and Khalid, M. S. 2017. Physico-Chemical Properties of Multi-Floral Honey from the West Bank, Palestine. *International Journal of Food Properties* 20(2): 447–454.

Ahmed, S., Sulaiman, S. A., Baig, A. A., Ibrahim, M., Liaqat, S., Fatima, S., Jabeen, S., Nighat Shamim, N., and Othman, N. H. 2018. Honey as a Potential Natural Antioxidant Medicine: An Insight into Its Molecular Mechanisms of Action. *Oxidative Medicine and Cellular Longevity* 2018: 1–19. https://doi.org/10.1155/2018/8367846.

Ajibola, A. 2015. Physico-Chemical and Physiological Values of Honey and Its Importance as a Functional Food. *International Journal of Food and Nutritional Sciences* 2(2): 180–188.

Alam, J. A., Fajloun, Z., Chabni, A., and Millet, M. 2017. The Use of Honey as Environmental Biomonitor of Pesticides Contamination in Northern Lebanon. *Euro-Mediterranean Journal of Environmental Integration* 2: 23.

Al-Rifai, J., and N. Akeel, N. 1997. Determination of Pesticide Residues in Imported and Locally Produced Honey in Jordan. *Journal of Apicultural Research* 36(3–4): 155–161.

Al-Waili, N., Salom, K., Al-Ghamdi, A., and Ansari, M. J. 2012. Antibiotic, Pesticide, and Microbial Contaminants of Honey: Human Health Hazards. *The Scientific World Journal* 2012: Article ID 930849.

Barhate, R. S., Subramanian, R., Nandini, K. E., and Umesh Hebbar, H. 2003. Processing of Honey using Polymeric Microfiltration and Ultrafiltration Membranes. *Journal of Food Engineering* 60: 49–54.

Bartha, S., Taut, I., Goji, G., Vlad, I. A. S., and Dinulica, F. 2020. Heavy Metal Content in Polyfloral Honey and Potential Health Risk. A Case Study of Cops, a Mica, Romania. *International Journal of Environmental Research and Public Health* 17: 1507. doi:10.3390/ijerph17051507.

Batinic, K. and Palinic, D. 2014. *Priručnik o medu.* Sveucilista u Mostaru, Agronomski i prehrambeno-tehnološki fakultet- Mostar, Federalni agromediteranski zavod Mostar.

Bertelli, D., Lolli, M., Papotti, G., Bortolotti, L., Serra, G., and Plessi, M. 2010. Detection of Honey Adulteration by Sugar Syrups Using One-Dimensional and Two Dimensional High-Resolution Nuclear Magnetic Resonance. *Journal of Agricultural and Food Chemistry* 58: 8495–8501.

Bhandari, B. R., Datta, N., and Howes, T. 1997. Problem Associated with Spray Drying of Sugar Rich Foods. *Drying Technology*15: 671–684.

Bhandari, B., D'Arcy, B., and Kelly, C. 1999. Rheology and Crystallization Kinetics of Honey: Present Status. *International Journal of Food Properties* 2: 217–226.

Blidi, S., Gotsiou, P., Loupassaki, S., Grigorakis, S., and Calokerinos, A. C. 2017. Effect of Thermal Treatment on the Quality of Honey Samples from Crete. *Advances in Food Science and Engineering* 1: 1.

Bogdanov, S. 1984. Honigdiastase, Gegenüberstellung verschiedener Bestimmungsmethoden. *Mitteilungen aus Gebiete Lebensmitteluntersuchung Hygiene* 75: 214–220.

Bogdanov, S. 2004. Contaminants of Bee Products. *Apidologie* 37 (2006): 1–18.

Bogdanov, S. 2011b. Physical properties. In S. Bogdanov (Ed.), The honey book (pp. 19–27). http://www.bee-hexagon.net/honey/ (Accessed Jan, 2021)

Bogdanov, S., and Martin, P. 2002. Honey Authenticity: A Review. *Mitteilungen aus dem Gebiete der Lebensmitteluntersuchung und Hygiene* 93: 232–254. Retrieved from http://www.bag.admin.ch/dokumentation/publikationen/02212/inex.html?lang=de.

Bogdanov, S., Baumann, S. E. 1988. Bestimmung von Honigzucker mit HPLC. *Mitteilungen aus Gebiete Lebensmitteluntersuchung Hygiene* 79: 198–206.

Bogdanov, S., Lullman, C., Martin, P., Von der Ohe, W., Russmann, H., Vorwohl, G., and Vit, P. 1999. Honey Quality and International Regulatory Standards: Review by the International Honey Commission. *Bee World* 80: 61–69. doi: 10.1080/0005772X.1999.11099428.

Borawska, M. H., Kapala, J., Hukalowicz, K., and Markiewicz, R. 2000. Radioactivity of Honeybee Honey. *Bulletin of Environmental Contamination and Toxicology* 64: 617–621.

Borylo, A., Romanczyk, G., Wieczorek, J., Parulska, D. S., and Kaczor, M. 2018. Radioactivity of Honey from Northern Poland. *Journal of Radioanalytical and Nuclear Chemistry* 319: 289–296.

Cavia, M. M., Fernandez-Muino, M. A., Huidobro, J. F., and Sancho, M. T. 2004. Correlation Between Moisture and Water Activity of Honeys Harvested in Different Years. *Journal of Food Science* 69: 368–370. doi:10.1111/j.1365-2621.2004.tb10699.x.

Chen, T. B., Deng, W. H., Lu, W. H., Chen, R. M., and Rao, P. F. 2001. Detection of Residual Antibiotics in Honey with Capillary electrophoresis. *School Equipment Production Unit* 19(1): 91–93.

Chin, N. K. and Sowndhararajan, K. 2020. A Review on Analytical Methods for Honey Classification, Identification and Authentication. In. *Honey Analysis and New Challenges*, ed. V. D. A. A. D Toledo, and E. D. Chambo, 1–33. Intech Open.

Choudhary, A. and Sharma, D. C. 2008. Pesticide Residues in Honey Samples from Himachal Pradesh (India). *Bulletin of Environmental Contamination and Toxicology* 80(5): 417–422.

Codex Alimentarius. 2001. Codex Alimentarius standard for honey 12-1981. Revised Codex standard for honey. Standards and standard methods (Vol. 11). http://www.codexalimentarius.net (Accessed Jan 10, 2021).

Conti, M. E., Stripeikis, J., Campanella, L., Cucina, D., and Tudino, M. B. 2007. Characterization of Italian Honeys (Marche Region) on the basis of their Mineral Content and Some Typical Quality Parameters. *Chemistry Central Journal* 1: 1–14.

Cordella, C., Militao, J. S., Clement, M.-C., Drajnudel, P., and Cabrol-Bass, D. 2005. Detection and Quantification of Honey Adulteration via Direct Incorporation of Sugar Syrups or Bee-Feeding: Preliminary Study Using High-Performance Anion Exchange Chromatography with Pulsed Amperometric Detection (HPAEC-PAD) and Chemometrics. *Analytica Chimica Acta* 531: 239–248.

Costa, P. A., Moraes, I. C. F., Bittante, A. M. Q. B., Sobral, P. J. A., Gomide, C. A., and Carrer, C. C. 2013. Physical Properties of Honeys Produced in the Northeast of Brazil. *International Journal of Food Studies* 2: 118–125. doi:10.7455/ijfs/2.1.2013.a9.

Cotte, J. F., Casabianca, H., Giroud, B., Albert, M., Lheritier, J., and Grenier-Loustalot, M. F. 2004. Characterization of Honey Amino Acid Profiles Using High-Pressure Liquid Chromatography to Control Authenticity. *Analytical and Bioanalytical Chemistry* 378: 1342–1350.

Crittenden, A. N. 2011. The Importance of Honey Consumption in Human Evolution. *Food and Foodways* 19(4): 257–273.

del Nozal, M. J., Bernal, J. L., Marinero, P., Diego, J. C., Frechilla, J. I., Higes, M., and Llorente, J. 1998. High Performance Liquid Chromatographic Determination of Organic Acids in Honeys from Different Botanical Origin. *Journal of Liquid Chromatography & Related Technologies* 21: 3197–3214.

Diaz-Basantes, M. F., Conesa, J. A., Fullana, A. 2020. Microplastics in Honey, Beer, Milk and Refreshments in Ecuador as Emerging Contaminants. *Sustainability* 2020(12): 5514.

Dimiņš, F., Miķelsone, V., Augspole, I., and Niklavs, A. 2019. Microwave Facilities for Thermal Treatment of Honey. *Key Engineering Material Zurich* 800: 103–107. doi:10.4028/www.scientific.net/KEM.800.103.

do Nascimento, S. A., Marchini, C. L., de Carvalho, L. A. C., Araújo, D. F. D., de Olinda, A. R., and da Silveira A. T. 2015. Physical-Chemical Parameters of Honey of Stingless Bee (Hymenoptera: Apidae). *American Chemical Science Journal* 7(3): 139–149.

Doner, L. W., White, J. W., and Phillips, J. G. 1979. Gas-Liquid Chromatographic Test for Honey Adulteration by High Fructose Corn Syrup. *Journal of the Association of Official Analytical Chemists* 62: 186–189.

European Chemical Agency. 2010. Evaluation of New Scientific Evidence Concerning the Restrictions Contained in Annex XVII to Regulation (EC) No. 1907/2006 (REACH): Review of New Available Information for Di-'isononyl'Phthalate (DIN); European Chemical Agency: Helsinki, Finland 2010.

European Council. 2001. Council Directive 2001/110/EC relating to honey. *Official Journal of the European Communities* 2002(L10): 47–52.

Fakhlaei, R., Jinap Selamat, J., Khatib, A., Razis, A. F. A., Sukor, R., Ahmad, S. and Babadi, A. A. 2020. The Toxic Impact of Honey Adulteration: A Review. *Foods* 9: 1538. doi:10.3390/foods9111538.

Fakhrildin, M. B., and Alsaadi, R. A. 2014. Honey Supplementation to Semen-Freezing Medium Improves Human Sperm Parameters Post-Thawing. *Journal of Family and Reproductive Health* 8 (1): 27–31.

Forsgren, E. 2010. European Foulbrood in Honey Bees. *Journal of Invertebrate Pathology* 103(1): S5–S9.

Ghazali, H. M., Ming, T. C., and Hashim, D. M. 1994. Effect of Microwave Heating on the Storage and Properties of Starfruit Honey. *ASEAN Food Journal* 9: 30–35.

Global Industry Analysts Inc. 2016. Honey: A Global Strategic Business Report. San Jose, CA: Global Industry Analysts. http://www.strategyr.com/MarketResearch/Honey_Market_Trends.asp (Accessed Jan 10, 2021).

Goharshenasan, P., Amini, S., Atria, A., Abtahi, H., and Khorasani, G. 2016. Application of Honey on Surgical Wounds: A Randomized Clinical Trial. *Forsch. Komplementmed* 23 (1): 12–15.

Gomez-Perez, M. L., Plaza-Bolanos, P., Romero-Gonzalez, R., Martınez-Vidal, J. L., and Garrido-Frenich, A. 2012. Comprehensive Qualitative and Quantitative Determination of Pesticides and Veterinary Drugs in Honey Using Liquid Chromatography–Orbitrap High Resolution Mass Spectrometry. *Journal of Chromatography A* 1248: 130–138.

Gonzalez Paramas, A. M., Barez, A. G., Garcia-Villanova, R. J., Pala, T. R., Albajar, A. R., and Sanchez, J. S. 2000. Geographical Discrimination of Honeys by Using Mineral Composition and Common Chemical Quality Parameters. *Journal of Science of Food and Agriculture* 80: 157–165.

Granja, R., Nino, A. M., Zucchetti, R., Nino, R. M., Patel, R., and Salerno, A. G. 2009. Determination of Erythromycin and Lyiosin Residues in Honey by LC/MS/MS. *Journal of AOAC International* 92(3): 975–980.

Handa, Y., Hirai, Y., Matsubara, T., and Sakurai, H. 1997. Radioactivity due to Several Radionuclides Detected in Honey of Different Geographical Origins. *American Bee Journal* 137: 307–309.

Haroun, M. I., Poyrazoglu, E. S., Konar, N., and Artik, N. 2012. Phenolic Acids and Flavonoids Profiles of Some Turkish Honeydew and Floral Honeys. *Journal of Food Technology* 10: 39–45. doi:10.3923/jftech.2012.39.45.

Hasan, S. H. 2013. Effect of Storage and Processing Temperatures on Honey Quality. *Journal of Babylon University/Pure and Applied Science* 21(6): 2013.

Hebbar, U. H., Nandini, K. E., Lakshmi, M. C., and Subramanian, R. 2003. Microwave and Infrared Heat Processing of Honey and Its Quality. *Food Science and Technology Research* 9: 49–53.

Iglesias, M. T., Martin-Alvarez, P. J., Polo, M. C., de Lorenzo, C., and Pueyo, E. 2006. Protein Analysis of Honeys by Fast Protein Liquid Chromatography: Application to Differentiate Floral and Honeydew Honeys. *Journal of Agricultural and Food Chemistry* 54: 8322–8327.

International Honey Commission. 2009. Harmonised methods of the International Honey Commission. *International Honey Commission* 1–66.

Irawati, N., Isa, N. M., Mohamed, A. F., Rahman, H. A., Harun, S. W., and Ahmad, H. 2017. Optical Microfiber Sensing of Adulterated Honey. *IEEE Sensors Journal* 17: 5510–5514.

Itoh, S., Yoshioka, K., Terakawa, M., Sekiguchi, Y., Kokubo, K. I., and Watanabe, A. 1999. The Use of Ultrafiltration Membrane Treated Honey in Food Processing. *Journal of Japanese Society for Food Sciences and Technology* (Nippon Shokuhin Kagaku Kogaku Kaishi) 46: 293–302.

Jakubik, A. P., Borawska, M. H., and Socha, K. 2020. Modern Methods for Assessing the Quality of Bee Honey and Botanical Origin Identification. *Food* 9: 1028.

Janghu, S., Bera, M. B., Nanda, V., and Rawson, A. 2017. Study on Power Ultrasound Optimization and Its Comparison with Conventional Thermal Processing for Treatment of Raw Honey. *Food Technology and Biotechnology* 55(4): 570–579.

Jovanovic, N. 2015. *Antimikrobna i antioksidativna aktivnost različitih uzoraka meda iz okoline Nisa*, Magistersi rad. Prirodno matematicki fakultet -Nis, Univerzitet u Nis.

Kale Sniderman, J. M., Matley, K. A., Haberle, S. G., and Cantrill, D. J. 2018. Pollen Analysis of Australian Honey. *PLOS ONE* 13(5): 1–24.

Karabagias, I. K., Vavoura, M. V., Nikolaou, C., Badeka, A., Kontakos, S., and Kontominas, M. G. 2014. Floral Authentication of Greek Unifloral Honeys based on the Combination of Phenolic Compounds, Physicochemical Parameters and Chemometrics. *Food Research International* 62: 753–760.

Kim, H. J., Lim, H. S., Lee, K. R., Choi, M. H., Kang, N. M., Lee, C. H., Oh, E. J., and Park, H. K. 2017. Determination of Trace Metal Levels in the General Population of Korea. *International Journal of Environmental Research and Public Health* 14: 702.

Kishore, R. K., Halim, A. S., Syazana, M. S. N., and Sirajudeen, K. N. S. 2011. Tualang Honey has Higher Phenolic Content and Greater Radical Scavenging Activity Compared With Other Honey Sources. *Nutrition Research* 31(4): 322–325.

Korta, E., Bakkali, A., Berrueta, L. A., Gallo, B., and Vicente, F. 2002. Study of an Accelerated Solvent Extraction Procedure for the Determination of Acaricide Residues in Honey by High Performance Liquid Chromatography-Diode Array Detector. *Journal of Food Protection* 65(1): 161–166.

Kowalczuk, I., Jezewska-Zychowicz, M., and Trafialek, J. 2017. Conditions of Honey Consumption in Selected Regions of Poland. *Acta Scientiarum. Polonorum Technologia Alimentaria* 16(1): 101–112.

Kuc, J., Grochowalski, A., Kostina, M. 2017. Determination of the Diastase Activity in Honeys. *Technical Transactions* 8/2017: 29–35.

Kumar, A., Gill, J. P. S., Bedi, J. S., Chhuneja, P. K., and Kumar, A. 2020. Determination of Antibiotic Residues in Indian Honeys and Assessment of Potential Risks to Consumers. *Journal of Apicultural Research* 59(1): 25–34.

Kumaravelu, C., and Gopal, A., 2015. Detection and Quantification of Adulteration in Honey Through Near Infrared Spectroscopy. *International Journal of Food Properties* 18(9): 1930–1935. doi: 10.1080/10942912.2014.919320.

Lamas, G. A., Navas-Acien, A., Mark, D. B., and Lee, K. L. 2016. Heavy Metals, Cardiovascular Disease, and the Unexpected Benefits of Chelation Therapy. *Journal of the American College of Cardiology* 67: 2411–2418.

Lane, J. H., and Eynon, L. 1923. *Journal of Society of Chemical Industry* 42: 32.

Lee, C. Y., Smith, N. L., Kime, R. W., and Morse, R. A. 1985. Source of the Honey Protein Responsible for Apple Juice Clarification. *Journal of Apicultural Research* 24: 190–194.

Li, S., Zhang, X., Shan, Y., Su, D., Ma, Q., Wen, R., and Li, J. 2017. Qualitative and Quantitative Detection of Honey Adulterated with High-Fructose Corn Syrup and Maltose Syrup by Using Near-Infrared Spectroscopy. *Food Chemistry* 218: 231–236.

Liebezeit, G., and Liebezeit, E. 2015. Origin of Synthetic Particles in Honeys. *Polish Journal of Food and Nutrition Sciences* 65(20): 143–147.

Lim, S., Cho, Y. M., Park, K. S., and Lee, H. K. 2010. Persistent Organic Pollutants, Mitochondrial Dysfunction, and Metabolic Syndrome. *Annals of the New York Academy of Sciences* 1201: 166–176.

Lorenz, E. S. 2009. Potential Health Effects of Pesticides. *Agricultural Communication and Marketing* 2009: 1–8.

Machado De-Melo, A. A., Almeida-Muradian, L. B. de., Sancho, M. T., and Maté, A. P. 2017. Composition and Properties of *Apis mellifera* Honey: A Review. *Journal of Apicultural Research* 1–33. https://doi.org/10.1080/00218839.2017.1338444.

Maghsoudi, H., and Moradi, S., 2015. Honey: A Skin Graft Fixator Convenient for Both Patient and Surgeon. *Indian Journal of Surgery* 77 (Suppl. 3): 863–867.

Mates, J. M., Segura, J. A., Alonso, F. J., and Marquez, J. 2010. Roles of Dioxins and Heavy Metals in Cancer and Neurological Diseases using ROS-Mediated Mechanisms. *Free Radical Biology and Medicine* 49: 1328–1341.

Mato, I., Huidobro, J. F., Simal-Lozano, J., and Sancho, M. T. 2003. Significance of Nonaromatic Organic Acids in Honey. *Journal of Food Protection* 66: 2371–2376.

Mejias, E., and Garrido, T. 2007. Analytical Procedures for Determining Heavy Metal Contents in Honey: A Bioindicator of Environmental Pollution. In *Honey Analysis*, ed. V. de Alencar and A. Toledo de, 311–324. INTECH: London, UK.

Meo, S. A., Al-Asiri, S. A., Mahesar, A. L., and Ansari, M. J. 2017. *Saudi Journal of Biological Sciences* 24: 975–978.

Nakano, H., Yoshikuni, Y., Hashimoto, H., and Sakaguchi, G. 1992. Detection of *Clostridium botulinum* in Natural Sweetening. *International Journal of Food Microbiology* 16(2): 117–121.

Nayik, G. A., Shah, T. R., Muzaffar, K., Wani, S. A., Gull, A., Majid, I. and Bhat, F. M. 2014. Honey: Its History and Religious Significance: A Review. *Universal Journal of Pharmacy* 03(01): 5–8.

Nevas, M., Hielm, S., Lindstrom, M., Horn, H., Koivulehto, K., and Korkeala, H. 2002. High Prevalence of *Clostridium botulinum* Types A and B in Honey Samples Detected by Polymerase Chain Reaction. *International Journal of Food Microbiology* 72(1–2): 45–52.

Noori, S., Al, W., Faiza, S. Al. W., Mohammed, A., Amjed, A., Khelod, Y. S., Ahmad, A., and Al, G. 2014. Effects of Natural Honey on Polymicrobial Culture of Various Human Pathogens. *Archives of Medical Sciences* 10(2): 246–250.

Olaitan, P. B., Adeleke, O. E., and Ola, I. O. 2007. Honey: A Reservoir for Microorganisms and an Inhibitory Agent for Microbes. *African Health Sciences* 7(3): 159–165.

Olivares-Pérez, A., Mejias-Brizuela, N., Grande-Grande, A., and Fuentes-Tapia, I. 2012. Corn Syrup Holograms. *Optik* 123: 447–450.

Ortelli, D., Edder, P., and Corvi, C. 2004. Analysis of Chloramphenicol Residues in Honey by Liquid Chromatography-Tandem Mass Spectrometry. *Chromatographia* 59(1–2): 61–64.

Ough, C. 1969. Rapid Determination of Proline in Grapes and Wines. *Journal of Food Science* 34: 228–230. (1969) DIN Norm 10754 (Entwurf 1991) Bestimmung des Prolingehalts von Honig.

Padovan, G. J., De Jong, D., Rodrigues, L. P., and Marchini, J. S. 2003. Detection of Adulteration of Commercial Honey Samples by the 13C/12C Isotopic Ratio. *Food Chemistry* 82: 633–636.

Pavlova, T., Dimov, I., and Nakov, G. 2018. Quality Characteristics of Honey: A Review. *Proceedings of University of Ruse* 57: book 10.2. University of Ruse "Angel Kanchev", Branch Razgrad, Bulgaria.

Perna, A., Intaglietta, I., Simonetti, A., and Gambacorta, E. 2013. A Comparative Study on Phenolic Profile, Vitamin C Content and Antioxidant Activity of Italian Honeys of Different Botanical Origin. *International Journal of Food Science and Technology* 48: 1899–1908.

Pourtallier, J., and Taliercio, Y. 1972. Honey Control Analyses. *Apiacta*, (1): 1–4.

Prodolliet, J., and Hischenhuber, C. 1998. Food Authentication by Carbohydrate Chromatography. *Zeitschrift für Lebensmitteluntersuchung und – Forschung A.* 207: 1–12.

Puscas, A., Hosu, A., and Cimpoiu, C. 2013. Application of a Newly Developed and Validated High-Performance Thin-Layer Chromatographic Method to Control Honey Adulteration. *Journal of Chromatography A* 1272: 132–135.

Rahman, M. M., Gan, S. H., and Khalil, M. I. 2014. Neurological Effects of Honey: Current and Future Prospects. *Evidence-Based Complementary and Alternative Medicine* 2014: 1–12. http://dx.doi.org/10.1155/2014/958721.

Ribeiro R. O. R., Marsico, E. T., Carneiro, C. S., Monteiro, M. L. G., Junior, C. A. C., Mano, S., and de Jesus E. F. O. 2014a. Classification of Brazilian Honeys by Physical and Chemical Analytical Methods and Low Field Nuclear Magnetic Resonance (LF 1HNMR). *LWT – Food Science and Technology* 55: 90–95.

Ribeiro, R. O. R., Marsico, E. T., Carneiro, C. S., Monteiro, M. L. G., Junior, C. A. C., and de Jesus E. F. O. 2014b. Detection of Honey Adulteration of High Fructose Corn Syrup by Low Field Nuclear Magnetic Resonance (LF 1H NMR). *Journal of Food Engineering* 135: 39–43.

Riswahyuli, Y., Rohman, A., Setyabudi, F. M. C. S., and Raharjo, S. 2020. Indonesian Wild Honey Authenticity Analysis Using Attenuated Total Reflectance-Fourier Transform Infrared (ATR-FTIR) Spectroscopy Combined with Multivariate Statistical Techniques. *Heliyon* 6 (2020): e03662.

Ritter, L., Solomon, K. R., Forget, J., Stemeroff, M., and Leary, C. O. 2007. Persistent organic pollutants: an assessment report on: DDT, Aldrin, Dieldrin, Endrin, Chlordane, Heptachlor, Hexachlorobenzene, Mirex, Toxaphene, Polychlorinated Biphenyls, Dioxins and Furans. Prepared for The International Programme on Chemical Safety (IPCS), within the framework of the Inter-Organization Programme for the Sound Management of Chemicals (IOMC) 2007.

Saito, K., and Tomita. 2000. Difructose Anhydrides: Their Mass-Production and Physiological Functions. *Bioscience Biotechnology Biochemistry* 64(7): 1321–1327.

Saridaki-Papakonstadinou, M., Andredakis, S., Burriel, A., and Tsachev, I. 2006. Determination of Tetracycline Residues in Greek Honey. *Trakia Journal of Sciences* 4(1): 33–36.

Saxena, S., Panicker, L., and Gautam, S. 2014. Rheology of Indian Honey: Effect of Temperature and Gamma Radiation. *International Journal of Food Science* 2014: 935129.

Schade, J. E., Marsh, G. L., and Eckert, J. E. 1958. Diastase Activity and Hydroxymethylfurfural in Honey and Their Usefulness in Detecting Heat Adulteration. *Food Research* 23: 446–463.

Schocken-Iturrino, R. P., Carneiro, M. C., Kato, E., Sorbara, J. O. B., Rossi, O. B., and Gerbasi, L. E. R. 1999. Study of the Presence of the Spores of *Clostridium botulinum* in Honey in Brazil. *FEMS Immunology and Medical Microbiology* 24(3): 379–382.

Shapla, U. M., Md. Solayman, M., Alam, N., Khalil, M. I., and Gan, S. H. 2018. 5-Hydroxymethylfurfural (HMF) Levels in Honey and Other Food Products: Effects on Bees and Human Health. *Chemistry Central Journal* 12(35): 1–18.

Siegenthaler, U. 1975. Bestimmung der Amylase in Bienenhonig mit einem handelsublichen, farbmarkierten Substrat. *Mitteilungen aus Gebiete Lebensmitteluntersuchung Hygiene* 66: 393–399.

Siegenthaler, U. 1977. Eine einfache und rasche Methode zur Bestimung der α-Glucosidase (Saccharase) im Honig. *Mitteilungen aus Gebiete Lebensmitteluntersuchung Hygiene* 68: 251–258.

Sivakesava, S., and Irudayaraj, J. 2000. Determination of Sugars in Aqueous Mixtures Using Mid-Infrared Spectroscopy. *Applied Engineering in Agriculture* 16(5): 543.

Snowdon, J. 1999. The Microbiology of Honey-Meeting Your Buyers' Specifications. *American Bee Journal* 139: 51–60.

Snowdon, J. A., and Cliver, D. O. 1996. Microorganisms in Honey. *International Journal of Food Microbiology* 31(1–3): 1–26.

Subramanian, R., Hebbar, H. U., and Rastogi, N. K. 2007. Processing of Honey: A Review. *International Journal of Food Properties* 10: 127–143.

Tchounwou, P. B., Yedjou, C. G., Patlolla, A. K., and Sutton, D. J. 2012. Heavy Metal Toxicity and the Environment. In *Molecular, Clinical and Environmental Toxicology*, ed. A. Luch, 134–164. Experientia Supplementum, Springer: Basel, Switzerland, 2012; Volume 101, pp. 134–164.

Testa, R., Asciuto, A., Schifani, G., Schimmenti, E., and Migliore, G. 2019. Quality Determinants and Effect of Therapeutic Properties in Honey Consumption. An Exploratory Study on Italian Consumers. *Agriculture* 9(174): 1–12. doi:10.3390/agriculture9080174.

Tette, P. A. S., Guidi, L. R., de Abreu Gloria, M. B., and Fernandes, C. 2016. Pesticides in Honey: A Review on Chromatographic Analytical Methods. *Talanta* 149: 124–141.

Thrasyvoulou, A., Manikis, J., and Tselios, D. 1994. Liquefying Crystallized Honey with Ultrasonic Waves. *Apidologie* 25: 297–302.

Thrasyvoulou, A., Tananaki, C., Goras, G., Karazafiris E., Maria Dimou, M., Liolios, V., Kanelis, D., and Gounari, S. 2018. Legislation of Honey Criteria and Standards. *Journal of Apicultural Research* 57(1): 88–96. https://doi.org/10.1080/00218839.2017.1411181.

Tosi, E. A., Re, E., Lucero, H., and Bulacio, L. 2004. Effect of Honey High-Temperature Short-Time Heating on Parameters Related to Quality, Crystallisation Phenomena and Fungal Inhibition. *Lebensmittel-Wissenschaft Und-Technologie* 37: 669–678.

Tura, A. G., Seboka, D. B. 2020. Review on Honey Adulteration and Detection of Adulterants in Honey. *International Journal of Gastroenterology* 4(1): 1–6. http://www.sciencepublishinggroup.com/j/ijg.

Veana, F., Flores-Gallegos, A. C., Gonzalez-Montemayor, A. M., Michel-Michel, M., Lopez-Lopez, L., Aguilar-Zarate, P., Ascacio-Valdes, J. A., and Rodriguez-Herrera, R. 2018. Invertase: An Enzyme with Importance in Confectionery Food Industry. In *Enzymes in food technology* Springer: Singapore 2018: 187–212.

Velicer, C. M., Heckbert, S. R., Lampe, J. W., Potter, J. D., Robertson, C. A., and Taplin, S. H. 2004. Antibiotic Use in Relation to the Risk of Breast Cancer. *The Journal of the American Medical Association* 291(7): 827–835.

Wakhle, D. M., and Phadke, R. P. 1995. Design for Honey Processing Unit Part I. *Indian Bee Journal* 57: 144–146.

Wakhle, D. M., Phadke, R. P., Pais, D. V. E., and Shakuntala Nair, K. 1996. Design for Honey Processing Unit Part II. *Indian Bee Journal* 58: 5–9.

Wakhle, D. M. 2002. Beekeeping Technology – Production, Characteristics and uses of Honey and other products. In *Perspectives in Indian Apiculture*, ed. R. C. Mishra and R. Garg, 152–185. India: Agrobios.

Webb, B. G. 2012. *The Book of Judges*. USA: William B. Eerdmans Publishing House Oak, USA.

White, J. W. 1978. The protein content of honey. *Journal of Apicultural Research* 17: 234–238.

Williams, I. H. 2002a. Cultivation of GM Crops in the EU, Farmland Biodiversity and Bees. *Bee World* 83: 119–133.

Williams, I. H. 2002b. The EU Regulatory Framework for GM Foods in Relation to Bee Products. *Bee World* 83: 78–87.

Xue, X, Wang, Q, Li, Y, Wu, L, Chen, L, Zhao, J., and Liu, F. 2013. 2-Acetylfuran-3-Glucopyranoside as a Novel Marker for the Detection of Honey Adulterated with Rice Syrup. *Journal of Agricultural and Food Chemistry* 61: 7488–7493.

Yavuz, H., Guler, G. O., Aktumsek, A., Cakmak, Y. S., and Ozparlak, H. 2010. Determination of Some Organochlorine Pesticide Residues in Honeys from Konya, Turkey. *Environmental Monitoring and Assessment* 168(1–4): 277–283.

Zabrodska, B. and Vorlova, L. 2014. Adulteration of Honey and Available Methods for Detection – A Review. *Acta Veterinaria Brno* 83: S85–S102.

Żmijewska, E., Teper, D., Linkiewicz, A., and Sowa, S. 2013. Pollen from Genetically Modified Plants in Honey– Problems with Quantification and Proper Labeling. *Journal of Apicultural Science* 57(2): 5–19.

4

Value Addition in Beehive Products with Special Reference to Honey

Rajesh Kumar
Himachal Pradesh University, Shimla, India

Bharti Sharma, Raksha Rani, Riya Sharma and Younis Ahmad
Career Point University, Hamirpur, India

CONTENTS

DOI: 10.1201/9781003175964-4

Introduction

The most significant apiculture products are honey, wax, propolis, royal jelly, pollen, venom. These products can be used in their natural state as produced by a bee. However, there are several uses of primary bee products, where these act as a part of another product. The value of secondary products enhances with the addition of bee products due to their quality and almost mystical reputation and properties. Hence, secondary products, which are made up of primary bee products, are known as value-added products. Various primary products of beekeeping still do not have any market until they are added further, usually used as value-added products. If there will be other products in primary products, the value of primary products may increase. Thus, profitability of many beekeeping operations is also increasing. Many traditionally used other artificial products substitute beekeeping products due to their better accessibility, low price and simpler managing. Hence, the products which contained one or more main products of beekeeping are all value-added products. Besides this, the conjunction of different honeybee products may enhance their significance beyond their peculiar natural value. The main product of apiculture is honey, both from a quantitative as well as economic perspective. Its traces were also found in ancient times and were the first animal product used by humankind. Looking into the historical perspectives, honey is used by human civilisation in one or another form since time immemorial. The use of honey as a source of food and one of the main components in cultural and religious ceremonies is old. It also has a unique position among nutraceutical foods due to naturally available sugars present in it, which are the only source of natural sweetener found worldwide (Cartland, 1970; Crane, 1980; Zwaenepoel, 1984). The similarities in cultural diversity have brought up multiple uses of honey to formulate many other products. Honey produced from various *Apis* species is an absorber, digester, nutritious and preservative agent in traditional Indian medical systems, viz., Unani, Siddha, Homeopathy and Ayurveda (Ediriweera and Premarathna, 2012). The properties of honey are easy digestibility, palatability, increasing appetite, purification of blood, antibacterial, antitussive, antifungal etc., making it the most promising food around the globe. It also proves helpful in various chronic diseases like that of heart, kidney, liver and lungs, as well as a proven remedy against sore throat, cough, cold and fever etc. Apart from this, honey has also been used in the cure of indigestion, smoothening of skin, ear/eye infections and the diagnosis of pregnancy.

Value Addition in Apiculture Products

For enhancing the product's value, it is important to formulate, produce and enhance value addition. A study on opting for beekeeping to prolong livelihoods indicates that the primary bee product has an extensive user preference and endorse sustainable livelihood for various small farmers and other urban occupants (Devkota, 2020). Primary products from beekeeping can be made into value-added products, making candles from bee wax or baking cakes and bread by using honey (http://www.fao.org/3/w0076e/w0076e03.htm). Adding value to the bee products increases more products to sell, and a farmer's income is also increasing. A study by Gibbs and Muirhead (1998) on the beekeeping industry in Australia signifies that the primary products utilised from the occupation are honey and beeswax (Figure 4.1).

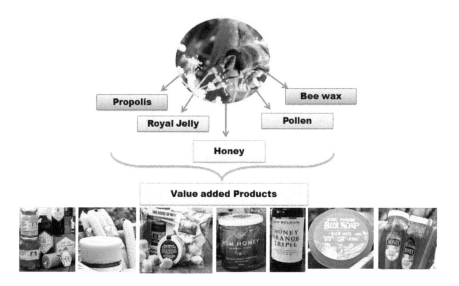

FIGURE 4.1 Showing various value-added products made up of beehive products.

Honey

Among the different wonders of nature, honey is one of them. Honey is known from ancient times, but yet we are a little familiar with it. It is the product of the nectar collected from the flowerings of various plants by bees. Nectar is placed inside the cells of a beehive and transformed by the worker bees adding different enzymes. Worker bees further eliminate the juicy part of nectar through fanning. The end product is heavy syrup consisting of 12–20% moisture and 80–85% sugar. It provides instant energy to the human body.

Honey is the valuable primary outcome of beekeeping, both quantitative and economical approach. It was also the first-ever product to be used by humanity during ancient times. The past of the honeybees goes parallel with the history of humankind, and its evidence can be seen in a various culture, using it as a food source and also seen as a figure engaged in religious, magic and therapeutic rituals etc. (Cartland, 1970; Crane, 1980; Zwaenepoel, 1984). The similar cultural fruitfulness has developed a similar colourful diversity of usage in other products. Honey is the natural sweetener developed by honeybees from the nectar of blossoms or from the discharge of an alive portion of plants or secretion of plants drawing insects upon the living part of plants, which is then collected, altered and combined with precise substances present with them by honeybees and stored and left in the honeycomb for ripening and maturing (Crane, 1975). Beside *Apis mellifera*, many other honeybee species are known for the production of honey and even other bees and wasps collects various kind of honey as their food reserves.

Properties of Honey

Viscosity

Freshly collected honey is an adhering liquid. The viscosity of honey is influenced by several constituents and, thus, differs with its composition and mainly depends on its water content. During honey processing, viscosity acts as an essential parameter, as it decreases the honey flow during extraction, pumping, settling, filtration, mixing and bottling. During the commercial processing of honey, the temperature of honey is elevated to lessen its viscosity. However, some honey display unlike features concerning viscosity, including 'Heather' (*Calluna vulgaris*), 'Manuka' (*Leptospermum scoparium*) and *Carvia callosa* are said to be thixotropic, which means that they acquire gel-like form when standing still and becomes liquid when disturbed. But contrary to this, eucalyptus honey varieties display completely contradictory features. Their viscosity elevates when stirred.

Density

Density is one of the important physical characteristics and is also known as specific gravity. It is more than the density of water and can also be determined by the water content of honey. Due to disparity in density, it is usually feasible to observe different stratification of honey in big storage containers. If the water content of honey is high, it will settle above the denser honey.

Hygroscopicity

For processing as well as final use, the intensely hygroscopic character of honey is significant. In value-added honey products such as pastry and bread, the propensity to absorb and maintain moisture is often desired. But, sometimes, the hygroscopicity becomes problematic due to excessive water content during processing or storage, which causes difficulty in preservation and storage. The honey with a water content of 18.3% or less will easily absorb the moisture from the atmosphere at a relative humidity beyond 60%.

Thermal Properties

Thermal property is an important attribute for honey processing. According to the composition and state of crystallisation of honey, its heat-absorbing capacity ranges from 0.56 to 0.73 calories per gram per degree Celsius, and its thermal conductivity differs from 118 to 143 × 10 Cal/cm^2/s/°C (White, 1975a). Thus, 'amount of heat', 'cooling' and 'mixing' can be calculated to treat a specific amount of honey before and after the pasteurisation. However, the low heat conductivity combined with high viscosity leads to fast overheating from the point heat source. Therefore, careful stirring is needed, and heating should be done only in water baths.

Colour

The colour of honey may differ from colourless to dark amber. Honey colours are all shades of yellow amber, such as the various dilutions of caramelised sugar used in ancient times as colour standards. The colour of honey depends upon types of flora, age and storage circumstances, but its transparency varies with the number of suspended particles like pollen. Honeys with less common colours bright yellow, reddish undertones, greyish and greenish are usually derived from sunflower, chestnut, eucalyptus and honeydew. When honey crystallises, its colour turns lighter due to the presence of glucose crystals (white-coloured). In some parts of East Africa, certain honey is found as white as milk that is wonderfully crystallised and is nearly water white in its liquid state.

The colour of honey is one of the essential features to assess its market value and use. Honey with light colour is sold for direct utilisation, whereas honey with darker colour is usually used for industries. In the global honey market, user demands are determined by honey's colour. Hence, after the quality confirmation, colour is the only vital factor deciding import and higher price. The honey colour is often specified on millimetres on the Pfund scale, which is usually an optical density reader used in international honey trade or conferring to US Department of Agriculture classifications (White, 1975c; Crane, 1980; Aubert and Gonnet, 1983; Rodriguez Lopez, 1985).

Crystallisation

Crystallisation is also an important feature for honey marketing but not for price purposes. In a moderate climate, most of the honey gets crystallised at usual storage conditions. This happens because the honey is an oversaturated sugar suspension that contains a high amount of sugar. Still, there is a misconception in consumers' minds that the crystallised honey has gone bad or mixed with sugar. The crystallisation of honey is due to the formation of 'monohydrate glucose crystals' that differs in quantity, form, dimensions and quality with the composition of honey and storage circumstances. If the water content is less and glucose content is high in honey, the rate of crystallisation is also higher. When the temperature is above 25°C and below 5°C, then virtually no crystallisation occurs. Although fast crystallisation occurs at the optimum temperature of 14°C, having solid particles such as pollen grain and slow stirring results in faster crystallisation. Hence, slow crystallisation creates larger and extra irregular crystals. During the process of crystallisation, water is released. Accordingly, the amount of water in the liquid phase increases, and the risk of fermentation also increase. Therefore, partially crystallised honey might cause problems in preservation. Thus, the need for controlled and absolute crystallisation is regularly induced intentionally. Moreover, partly crystallised honey is not an appealing giving for retail shelves. Main properties commonly ascribed to honey are concisely described above. According to specific cultures and traditions, there are many local uses of honey globally, and it is not possible to mention all (Donadieu, 1983). Several uses of honey and other primary products were mentioned in Koran also (EI-Banby, 1987).

Health and Medicinal Benefits of Honey

Honey is believed a miraculous natural product that facilitates better physical health and opposes fatigue, and it also supports better mental efficacy. It can be used by healthy and sick for any faintness, mainly about digestive and assimilative problems. Honey can improve the growth of a newborn baby who has given no breastfeeding and can improve the calcium fixation of bones; curing anaemia and anorexia might be attributed to nutritional benefits for eating honey.

Benefits to the Digestive Apparatus

Honey improves food absorption and benefits chronic and infectious intestinal problems, including duodenal ulcers, constipation and liver disturbance. It has been reported that honey can treat several gastrointestinal disorders (Salem, 1981; Haffejee and Moosa, 1985).

Benefits to the Respiratory System

Infections, cold and bronchial or throat irritations caused due to climate change and temperature fluctuation, honey is proved as a worldwide remedy against such things. The benefits of honey other than antibacterial effects, its fructose content is also assumed as soothing and relaxing.

Benefits to Skin and Wound Healing

Honey is used as a cosmetic product in conditioning and nourishing and in pharmaceutical preparations that can be applied directly on open injuries, lesions, bedsores, several ulcers and blisters. It is beneficial against infections, promotes tissue redevelopment and decreases scarring in its unprocessed or pure form (Hutton, 1966; Majno, 1975; Armon, 1980; Dumronglert, 1983). Honey reduces the scorching of burns and helps in the rapid generation of new tissue. A study revealed that a cream prepared from equal parts of honey, rye flour and olive oil was applied three times per day and positively used on various sores and open wounds, even on infected wounds in horses (Ibrahim et al., 2018). An effective test was done by Lucke (1935) in which honey and cod liver oil combination put off in plain unreactive cream base on open injuries in humans, but he gave no detailed-on proportion. Lücke (1935) carried out an effective test in which honey and cod liver oil combined with an unreactive cream base on open injuries of humans.

Benefit to Eye Disorders

It has been claimed through many clinical and traditional cases, which came from Europe (Mikhailov, 1950), Asia and Central America, that honey is found to be effective in reducing and treating eye cataracts, conjunctivitis and numerous afflictions of the cornea when used directly on the eye. Honey from Meliponoid and Trignoid honeybees inhabiting South and Central America and India are found to be the most effective. There are many cases where pure honey or a 3% sulphide ointment substituted by honey in place of Vaseline helped treat Ceratitis rosacea and corneal ulcers.

Medicine-Like Benefit

Specific advantages of uni-floral honey have been described regularly, depending upon the traditional supposition that honey synthesised from the nectar of medicinal plant possess same or identical favourable activity as the one acknowledged for the whole plants or some parts of it. Mechanisms identical to homeopathic potentiation are feasible, even when there is no transportation of active ingredients. Analytically efficient therapies, including Bach flower therapy and aromatherapy, indicate that honey's medicinal value is not restricted to the chemical analysis and quantification reveals. These claims are, however, not sustained by conventional scientific evidence.

Diabetes

There are many claims which believe that honey is beneficial for people with diabetes. It is unlikely to be approved due to its high sugar concentrations. Despite this, it is considered better than products synthesised from cane sugar (Katsilambros et al., 1988). It has been discovered that the level of insulin is low compared to the intake of equal caloric values of other foods. Still, blood sugar remained equal or higher than in the other compared products shortly after eating. In a healthy individual, the uptake of honey maintains a low blood sugar level compared to the intake of the same amount of sucrose (Shambaugh et al., 1990).

FIGURE 4.2   Showing various benefits of honey.

Ayurvedic Medicine

In Ayurvedic medicine of India, the honey is used chiefly for the absorption of various drugs such as herbal extracts. Beside this, it is also considered effective in treating various specific ailments, such as those linked to respiratory irritations and infections, mouth sores and eye cataracts. These are also used as general restoratives by newborn infants, the young and the elderly, the convalescent and hard-working farmers. At large, no differences have been marked between honeys from *A. mellifera*, *Apis cerana* or *Apis dorsata* (Figure 4.2).

Scientific Evidence

Various scientific studies have reported that honey should be considered food rather than medicine (Samarghandian et al., 2017). Most of the advantages depicted above, especially for internal usage, can provide a nutritional effect of some kind. On the other hand, present scientific knowledge suggested that any treatment is approved when a single compound is approved is found to be capable of affecting a well-defined symptom. However, there is no clarification of this fact when different compounds work synergistically or induce complex reactions.

Energy Source

The composition of honey is simple sugars such as fructose and glucose, which decide about all the indications on how, when and why to use it. Honey is an instant supplier of energy, which can be used as an energy booster by healthy and sick people: immediate access to energy without wasting much time digestion. In contrast, it also carries a threat to pathological sugar metabolisms, such as diabetes and obesity.

Non-Energetic Nutrients

Honey is often suggested to any age group due to nutrients like vitamins and minerals in it. Still, their low content makes it impractical to think that they can deliver enough essential supplements in the undersupplied diet. Similar suggestions have been made regarding the nutritional and health advantages of other bee products such as pollen and royal jelly. Even though the helpful features have been displayed in various cases, they cannot be based on the simple numeric value, i.e., X amount of substance Y. Still, the effectiveness of nutrients in the body is dependent on the quality and accessibility. Micronutrients present in unprocessed honey are believed to be of the topmost possible quality. From a nutritional perspective, a synergistic balancing effect or one that unravels the access to other nutrients already present is one of the more reasonable yet unconfirmed hypotheses.

Topical Applications

Topical usage under regulated conditions shows enhanced wound healing in animals (Bergman et al., 1983; El-Banby et al., 1989) and burn wounds caused by experiments in rats (Burlando, 1978) and is also useful in treating numerous types of wounds such as post-operative ones in humans (Cavanagh et al., 1971; Kandil et al., 1987a, 1987b, 1989; Efem, 1988; Green, 1988). Identical but not the same results were obtained using purified sucrose and special polysaccharide powders (Chirife et al., 1982). Post-operational wounds become bacteriologically disinfected within few days and dry out. If honey is applied simultaneously to wound, it stimulates tissue generation and reduces the time of healing. Moreover, bandages applied with honey do not adhere to the injuries or sensitive new skins. In hilly places where there are meagre medical facilities, honey has been employed effectively to treat various ailments since historic times.

Antibacterial Activity

The antibacterial activity of honey is the most studied activity of honey and simplest to test. The pH of honey ranges from 3.5 to 5.0, which means it, is acidic and its sugar concentration is high. However, diluted honey has been shown antibacterial activity, and an exclusive substance was derived known as inhibin. This activity was also later attributed to hydrogen peroxide during the formation of gluconic acid from glucose. The glucose oxidase enzyme is usually inactive in concentrated pure honey. Hence, in diluted honey with correct pH, antibacterial activity is mostly due to hydrogen peroxide. The primary significance of such a mechanism result from the need to protect unripe honey contained by colony till higher sugar concentration is attained.

Heggers et al. (1987) state that both mechanisms explain the sterilising effect of honey on injuries and some of its effectiveness against cold infections, but it does not clarify the useful effect of honey on burn injuries. Also, it does not explain quicker wounds with less damaged tissue. It has been experienced by Subrahmanyam (1993) that honey can help in skin graft after storage for about 12 weeks with 100% acceptance. Depending upon the types of honey, the antibacterial activity of honey is also varying (Dustmann, 1979; Revathy and Banerji, 1980; Jeddar et al., 1985; Molan et al., 1988). Furthermore, other unknown substances with antibacterial effects seem to contain honey, including polyphenols, to glucose oxidase. These factors have been recognised in some cases (Toth et al., 1987; Bogdanov, 1989; Molan et al., 1989), although as a whole, some scientific analysis is there which claims the useful effects of honey. On the other hand, it has been demonstrated that the maximum of the anti-bacterial activities of honey is lost after heating treatment or prolonged exposure to sunlight (Dustmann, 1979).

Information Sources on Honey Therapy

A book on honey therapy (in Rumanian) was published by Mladenov in Rumania, and several articles were published in Apimondia (1976) and Crane (1975, 1990). There is an American Apitherapy Society that collects the case histories and scientific information regarding all the therapeutical utilisation of beehive products (Krell, 1996).

Value-Added Products

The honeybee is a boon to human kinds that provides several useful products like honey, wax, propolis and royal jelly. Many of the beehive products can be used in their natural form, while there are many additional uses of bee products and can be used as part of other products. There are many value-added products of beekeeping which are also known as secondary products.

Mead

Mead is honey wine, and its quality and flavour depend on fermentation control and qualities of several ingredients, primarily on the features and flavour of the preferred honey. Better-quality honey with preferred savour should be selected, and a good water supply should be maintained for quality preparation. The water can control the flavour of the mead, mainly as the community water supplies have plenty of minerals, chemicals and other useful elements in them. The water of the well and soft rain is best but needs to be boiled initially.

The quantities/ratios of ingredients to make the mead mainly depend upon the water content of honey required sugariness and its alcohol content. Usually, in the final product, 2.3-kg honey/100 L of water for each alcohol grade is considered. More accurately, 21% of solid sugar must be added to get dry mead with 12% alcohol. When the quantity of solid sugar is increased about 25%, then it leads to a final alcohol content of 14–15%. More sugar gives remaining sugar in the ultimate product, so sugary mead is produced (Panesar et al., 2017). Pasteurisation is an important task in the preparation of the mead. Generally, it is not compulsory before fermentation. However, filtration is recommended to remove any solid granules. There is a school of mead makers who recommended pasteurisation (sterilisation) before the addition of preferred yeast. The pasteurisation could be attained either by adding tablets that produce SO_2 (sulphur dioxide), as utilised in ordinary wine makings, or by heating the mead at 78°C for 7 minutes. 'Bisulphite' and 'Campden' are the tablets that are used in pasteurisation. The sulphur dioxide gas does not flavour the mead, because it escapes during the process. These tablets are also used to sterilise bottles, corks, funnels and siphons.

Various salts and minerals, including urea, ammonium, phosphate, cream of tartar, tartaric and citric acid, are added to the chilled must as yeast nutrients. The acid improves the taste of the mead and inhibits the growth of unwanted microorganisms. Many recipes of nutrient combinations are listed below. Since fruit juice can provide nutrients and the right yeast, it can replace 50% of water, and there is no need for additives.

An active champagne yeast or brewer yeast of sufficient amount (0.5–2%) is added to the mead. The flavour of the mead is influenced by yeast choice, but its selection is more significant, aiming toward absolute and continuous fermentation. There is a need for actively growing yeast solutions that should be prepared for considerable groups. Yeast can be added directly to must if there are small batches. Qureshi and Tamhane (1985) disabled the yeast cell in calcium alginate cell to accelerate the fermentation process in mead production. To improve the taste, it must need flash heating of about 30 s to 102°C and immediate cooling to 7°C before adding the yeast (Kime et al., 1991).

It is beneficial to oxygenate the mead by transferring to another container after 2–3 days, however not compulsorily. Additionally, decanting is also useful to stop the bitter flavour coming from dead yeast gathered at the container base when the process of fermentation is longer. Then, the must should be remained undisturbed for about one month. Without stirring the sediment, the mead should be transferred with a hosepipe. This poring is not sufficient to purify the mead, which is made from honey. The addition of precipitating mediators, including tannins, bentonite colloidal protein solution or egg white, is necessary for the complete refining of the mead. Thus, the liquid gets filtered after few days. On the other hand, if it is boiled before the fermentation, it will remove the various proteins that are liable for clouding mead. But most of the honey aroma will also eliminate (Krell, 1996).

Honey Beer

In comparison to the mead, the process of making honey beer is simpler and quicker. However, honey beer cannot be stored for a longer time. Hence, it must be consumed within few hours. Nonetheless, once it has become uniform, it may be restored with the addition of more honey. There are various methods of making this widespread honey beer across the African continents.

Paterson in Crane described an exemplary profitable honey beer in Kenya containing substantial amount of 'refined cane sugar', 'jaggery or freshly squeezed cane juice'. If there is larger content of honey in the beer, it is considered better. Paterson revealed a recipe for making honey beer from 27 kg of honey with 108 kg of sugar in 250 L of water. In a study, 5 L of honey in 18 L of water were utilised by Kihwele from Dar-es-Salaam, Tanzania, mixed with six teaspoons of waterless yeast. The fermentation process should take place in the dark, so that consumption becomes ready after 5–7 days (Morse and Steinkraus, 1975). If the starting amount of the yeast is higher, the honey beer becomes readily available for consumption (within 1–2 days). However, more yeast (10 teaspoons of yeast per litre) can leave a sturdy flavour of yeast in the honey beer. When a small amount of yeast is added to the beer, it doesn't get ready to drink within 24 hours. Although, in Zambia, an author has seen a honey beer prepared within 6 hours in which yeast was added in the higher amount (Krell, 1996).

According to Morse and Steinkraus (1975), the temperature during the process of fermentation should be between 20 and 25°C. The fermentation process takes place at different temperatures and different speeds; hence, the accurate temperature is not necessary. Whenever the fermentation process is longer, the risk of infection by other yeasts or bacteria becomes high. Fermentation is faster at higher temperatures, but the production of the alcohol is also lower. Additionally, fermentation is lower at lower temperatures and ultimately stops.

Honey with Fruits and Nuts

Fruits in Honey

Various types of dry fruits, which have little moistness and softness, can be used with honey. They can be used in any form like whole, chopped and poured in the honey. Fruits with high moisture content or partially dry should be covered with honey for some days in air-tight containers to completely dry. The above process can be repeated (2–3 times) after the honey pouring until the honey gets no longer thinned with liquid juice coming from fruits. Afterwards, these fruits can be blended with an absolute batch of honey. Hence, this process is compulsory because fruit juice can add excessive water to honey that may spoil it. The pasteurisation of fruit and honey can enhance hygiene and preservability, leading to the possibility of fermentation, but the flavour of the honey may get affected. The remaining diluted pasteurised honey can be used in the form of fruit syrup (Krell, 1996).

Nuts in Honey

As the commercial nuts are already dried, they need not any further drying. The flavour of honey should be well mixed with selected nuts. Different types of glass jars can be used to store the mixture of honey and nuts. Honey and nuts should be mixed before bottling. Sometimes, an accurate amount of honey should be poured before the bottles and nuts mixed later. The ratio should be accurate for each nut type. Nuts should be packed tightly to not float on top and leave a pure honey layer at the bottom (Krell, 1996).

Honey Pastes for Dressing Wounds

Various types of ointments and cream are mixed with honey which is beneficial to dress injuries. These home remedies are multipurpose and used in the treatment of various ailments. Uccusic (1983) used recipe *viz.* wax (10 g), propolis extract, which is prepared in 10% ethanol solution (3 g), and honey (2 g). The wax should be melted first, and after cooling of wax, it is mixed with the propolis extract, and honey is added to it. Place the mixer in an airtight container and store it in a dark and cool place. Finally, this paste is ready to use. It can be used on all kinds of injuries, sores, wounds and infections, including paradentosis (damage of skin due to radiation), acids burn and poisonings (Krell, 1996).

Honey Jelly

There is one more useful product made from honey which is honey jelly. A recipe of honey jelly is in use in Unipectina Spa in Bergamo, Italy. To produce 1 kg honey jelly, water (220 ml), pectin (3–4 g), honey (800 g) and tartaric acid (1.5–2 ml) are used. In the first step, pectin needs to be immersed in the cold water and spread by rousing. Pectin is boiled until its weight decreases to 200 mg. Afterwards, honey is added, and the mixture is heated at a temperature of 60°C. After heating, remaining constituents, i.e., acid, are added, and the mixture is poured into the jar or container.

　If there is no availability of a mechanical mixture, dispersion of the pectin can be done in a little amount of honey, and also water can be added to the mixture. By heating the mixture at a temperature of 70°C, fermentation of the mixture can be avoided, and paste can pack without sterilisation. The packing jar needs to sterilise if the mixture is heated at 60–65°C. The percentage of final content should be 65–68%, with pH 3.1–3.3. Honey in the mixture serves as both sweetening and flavouring agent, and some parts of honey can be replaced with that of fruit juice or pulps to provide different flavours (Krell, 1996).

Candy

Candies and caramels provide various products in colour, consistency, flavours and shape, and it is an art by itself. The temperature is the main criteria for the consistency of the candy; therefore, a candy thermometer should be there at the time of heating. For the process of caramelisation, honey needs a higher temperature while adding in a recipe. Hence, another test can be utilised for the assessment of the right temperature. The critical temperature of the recipe can be recognised by colour, boiling behaving, threading of candy and adequate experience. While testing, hot and cool water should be used, and heat can be removed by avoiding overheating during the test. Rombauer and Rombauer Becker (1975) give descriptions of various stages of candy processing and caramelisation.

Honey Caramels

After Paillon (1960), ingredients used for making honey caramels are 0.75 g of honey, 6 g of sugar, 0.75 g of glucose, 2 ml of warm water and sufficient vanilla powder and alcohol. The first step in the preparation of honey caramel is heating the water in an odourless frying pan. Sugar is added to the frying pan while slowing down the heat. The mixture is stirred to evade the caramelisation on the base. Afterwards, glucose was added to the mid of the syrup. However, honey can be added to the place of syrup at a later stage. Now, the mixture was boiled for a while. The layer of foam was removed, and crystals were cleaned from the sides of the pan by wrapping it for nearly 3 minutes. Pan was then uncovered, stirred and heated at about 125–128°C temperature until the mixture became hard balls. At this stage, honey is added and heated continuously until the temperature reaches 145°C, a soft crack stage. Now, heating is stopped, and the material is transferred to the cold and oiled tray. The mixture is allowed to cool until it acquires pliability. The mixture is then spread consistently and squeezed out in the desired shapes. It is allowed to cool for a while, and sugar powder is added before the packing. Various flavours other than honey can be used in the caramels like mint, vanilla and eucalyptus. The cutting of the caramel must be done just after it becomes hard (Krell, 1996).

Honey-Roasted Nut Bars

This recipe is very malleable or resilient. The ratio of all the ingredients (sugar, honey and nuts) can be diverse than the caramel bar with few nuts that may be solid or caramelised with sugar and honey (nuts coated). The moisture-resistant packing material and cost consideration are the main criteria that determine the ratio of honey. Paillon (1960) used different ingredients in the preparation of honey-roasted nut bars like sugar (10 g), honey(2.5 g), almond and other nuts (whole or broken) (1.25 g), water (5 ml), white vinegar (1.25 ml). In the first step, sugar and vinegar are added to the water. The mixture is kept over the flame and stirred uninterruptedly. On boiling, honey was added and blended, and then the mixture was heated again to boil. Afterwards, the mixture was covered nearly for 3 minutes to remove the crystal from the edges of the pan. Pan was uncovered, and without stirring mixture was brought to a golden brown soft stage. Now, nuts are poured, and the mixture is baked for a while without increasing the temperature. The mixture is poured onto the light, greased or buttery tray. Then candies of different shapes are obtained before it becomes hard, and then these are wrapped in a moisture-proof package after cooling (Krell, 1996).

Honey in Bakery Products

Bread

Crane (1980) gave different ingredients used in the preparation of bread, including wheat flour (700 g), milk (450 ml), honey (7 ml), fresh yeast (20 g) and salt (5 g). Honey is added to the yeast and warm water, mixed and left for nearly 10 minutes. Now flour, salt and milk are added to the mixture to make its dough smooth and flexible. Flour is placed in a deep, slippery and pre-covered container for two and half hours to rise in a warm place. After that, the flour is separated in two, knead softly and left for 10 minutes to make the loaves of the dough and covered with the cloth. Loaves are allowed to dry in a warm place for an hour. Now, loaves are baked in preheated oven at a temperature of 220°C for nearly 40 minutes. However, if the baking soda is added instead of yeast, the recipe becomes easier, and there is no need for rising temperature in the dough (Krell, 1996).

Coconut Oat Cookies

Crane (1980) gave various ingredients used in the preparation of coconut oat cookies, including margarine (25 g), honey (4.5 g), flour (30 g), rolled oats (25 g), dries, shredded coconut (20 g), brown sugar (35 g), sodium bicarbonate (0.4 g) and warm water

(3 ml). Baking soda is mixed in water along with all the ingredients thoroughly. The mixture is then liquefied with margarine, and honey is added to it. All the ingredients were mixed in the bowl, poured into the greased baking sheet and baked at the temperature of 180°C for 10–15 minutes.

Honey Biscuits

Various ingredients used in the making of honey biscuits include flour (3.5 g), honey (1.2 g), and rolled oats (25 g), eggs (6/kg), baking powder (0.1 g) and warm water (3 ml). To prepare a honey biscuit, butter is heated and dissolved with honey. All the ingredients are added slowly to the bowl. Before making the loaves, the dough is cooled. Then it is cut down in the desired shapes of biscuit and baked in preheated oven for about 15 minutes at the temperature of 200°C (Krell, 1996).

Propolis

Propolis is a well-known honeybee product having complex chemical composition. Usually, it is made of sticky and balsamic material gathered by honeybees from flower buds, trees, shoots and other resinous exudates of plants. Honeybees use propolis to protect their hive from microorganisms, seal cracks, sterilise queen bee seats and wrap them with enemy invaders. In about 300 B.C., propolis is treated as a wonderful therapeutic used in many medicines worldwide (Bankova et al., 1988). Silva-Carvalho et al. (2015) revealed that this product is one of the natural drugs used from ancient times by different cultures. Many products are made from propolis, such as shampoos, antiseptic, chocolates, candies, lotion and toothpaste, which are being commercialised globally (Ramos and Miranda, 2007).

Medicinal Use of Propolis

Propolis is mainly used as an immunity booster (Wagh, 2013) and in healing infections. Its antibiotic activity blocks out the entry of viruses, bacteria and other organisms into the hive. It has been studied that propolis has some antioxidant, liver defending, anti-inflammatory (Daleprane and Abdalla, 2013) and anticancer properties. It has been used in the field of dermatology for the treatment of leg sores, neurodermatitis and dermatosis. However, propolis has also been used in advanced pulmonary tuberculosis and non-specific bronchitis, especially when the customary therapy has failed.

Bee Pollen

Bee pollen is one of the essential products of beekeeping that serve as natural food and an enhancer of good health and has been used worldwide. There is no other product on the earth that can compete with the nutritional value of bee pollen. Bee pollen is enriched with natural components, which are significantly important for the human body's metabolic activity and well-being (Komosinska-Vassev et al., 2015). Bee pollen is frequently called the 'only perfectly complete food'. Athletes are suggested to use pollen as their diet, which is a miracle food. Honeybees are served as role models for their activeness and are highly productive members of society. Generally, pollen grains are the male reproductive part of the flowers, formed in the anthers. Bee pollen has been used worldwide in various food or medicines. Bee collects nearly millions of pollen grains per flower and places them into pollen baskets with the help of special hairs that are situated on their hind legs. The gathered pollen is usually mixed with nectar to make it sticky to adhere to their legs during collection. Finally, these pollen pellets are harvested from the bee colony, which is sweet.

Pills and Capsules

Different types of pills and capsules are manufactured by processing the pollen. To process the pollen, a simple machine is required, which may even be second-hand. A mixture of honey and pollen is made to suppress. By layering the pills into wax, they can be made non-allergic, avoiding the link with mucous membranes. If there is no availability of pills, additional gum Arabic or gel and wax may be used to use individually. Encapsulation of dried pollen is the feasible and economical way for human consumption (Linskens and Jorde, 1997).

REFERENCES

Apimondia 1976. *Apitherapy today.* Apimondia Publishing House, Bucharest, Romania, 105 pp.

Armon, P.J. 1980. *The use of honey in the treatment of infected wounds.* Tropical Doctor, 10: 91.

Aubert, S. and Gonnet, M. 1983. *Mesure de la couleur des miels.* Apidologie, 14(2): 105–118.

Bankova, V., Popov, S., Marekov, N., Manolova, N. and Maksimova, V., 1988. *The chemical composition of propolis fractions with antiviral action.* Acta Microbiol. Bulgarica, 23: 52–57.

Bergman, A., Yanai, J., Weiss, I., Bell, D. and Menachem, P.D. 1983. *Acceleration of wound healing by topical application of honey. An animal model.* Am. J. Surg., 145: 374–376.

Bogdanov, 5. 1989. *Determination of pinocembrin in honey using HPLC.* I. Apic. Res., 28 (1): 55–57

Burlando, F. 1978. *About the therapeutic action of honey on burn wounds.* Minerva Dermatol., 113: 699–706

Cartland, B. 1970. *The magic of honey.* Corgi Books, London, UK, 160 pp.

Cavanagh, D., Beazley, J. and Ostapowicz, F., 1971. *Radical operation for carcinoma of the vulva. A new approach to wound healing. Obstet. Gynecol. Surv.,* 26(6): 460–461.

Chirife, J., Scarmato, G. and Herszage, L. 1982. *Scientific basis for use of granulated sugar in treatment of infected wounds.* The Lancet, 319, 560–561.

Crane, B. 1980. *A book of honey.* Oxford University Press, Oxford, U.K., 198 pp.

Crane, E. 1975. *Honey: A comprehensive survey.* Heinemann (in coop. with IBRA), London, U.K. 608 pp.

Crane, E. 1990. *Bees and beekeeping: Science, practice and world resources.* Cornstock Publ., Ithaca, NY. 593 pp

Daleprane, J.B. and Abdalla, D.S., 2013. *Emerging roles of propolis: antioxidant, cardio-protective, and antiangiogenic actions.* Evid. Complement. Altern. Med. 8. Article ID 175135, http://dx.doi.org/10.1155/2013/175135.

Devkota, K. 2020. Beekeeping: Sustainable livelihoods and agriculture production in Nepal. In: *In modern beekeeping-bases for sustainable production.* Intech Open. https://www.intechopen.com/chapters/70613.

Donadieu, Y. 1983. *Honey in natural therapeutics.* Maloine Editeur, S.A., Paris, 28 (2nd.)

Dumronglert, E. 1983. *A follow-up study of chronic wound healing dressing with pure natural honey.* J. Nat. Res. Council, Thailand, 15: 39–66

Dustmann, J.H. 1979. *Zurantibakteriellen Wirkung des Honigs.* Apiacta, 14: 7–11

Ediriweera, E.R.H.S.S. and Premarathna, N.Y.S. 2012. *Medicinal and cosmetic uses of Bee's Honey—A review.* Ayu., 33(2), p. 178.

Efem, S.E.E. 1988. *Clinical observations on the wound healing properties of honey.* Br. J. Surg., 75: 679–681

EI-Banby, M.A. 1987. *Honeybees in the Koran and in medicine.* Al-Ahram Centre for Translation and Publication, Cairo, Egypt, 205 pp.

El-Banby, M.A. et al. 1989. Healing effect of floral honey and honey from sugar-fed bees on surgical wounds (animal model). In: *Proceedings of 4th International Conference on Apiculture in Tropical Climates, Cairo, Egypt.* 6–10 November 1988. IBR, London, U.K. pp. 46–49.

Gibbs, D.M. and Muirhead, I.F. 1998. *The economic value and environmental impact of the Australian beekeeping industry.* Report prepared for the Australian Beekeeping Industry.

Green, A.E. 1988. *Wound healing properties of honey.* Br. J. Surg., 75(12): 1278

Haffejee, I.E. and Moosa, A. 1985. *Honey in the treatment of infantile gastroenteritis.* Br. Med. J. 290: 1866–1867

Heggers, J.P., Velanovich, V., Robson, M.C., Zoellner, S.M., Schileru, R.O.D.I.C.A., Boertman, J.A.N.E. and Niu, X.T., 1987. Control of burn wound sepsis: *a comparison of in vitro topical antimicrobial assays.* J. Trauma, 27(2): 176–179.

Hutton, D.J. 1966. *Treatment of pressure sores.* Nursing Times, 18: 1533–1534.

Ibrahim, N.I., Wong, S.K., Mohamed, I.N., Mohamed, N., Chin, K.Y., Ima-Nirwana, S., and Shuid, A.N. 2018. *Wound healing properties of selected natural products.* Int. J. Environ. Res. Public Health, *15*: 2360.

Jeddar, A., Kharsany, A., Ramsaroop, U.G., Bhamjee, A., Haffejee, I.E. and Moosa, A., 1985. *The antibacterial action of honey. An in vitro study.* South Afr. Med. J., 67(7): 257–258.

Kandil, A., El-Banby, M., Abdel-Wahed, K. AbouSehly, G. and Ezzat, N. 1987a. *Healing effect of true floral and false nonfloral honey on medical wounds.* J. Drug Res. (Egypt), 17(1–2): 71–75

Kandil, A., El-Banby, M.A., Abdel-Wahed, K., Abdel-Gawaad, M. and Fayez, M. 1989. *Curative properties of floral honey and honey from sugar-fed bees on induced gastric ulcers.* Proc. 4th Intern. Conf. Apic. Trop. Clim., 68–69.

Kandil, A., El-Bandy, M., Abdel-Wahed, K., Abdel-Gawaad, M. and Fayez, M. 1987b. *Curative properties of true floral and false non-floral honeys on induced gastric ulcers.* J. Drug Res. Egypt, 17(1–2): 103–106.

Katsilambros, N.L., Philippides, P., Touliatou, A., Georgakopoulos, K., Kofotzouli, L., Frangaki, D., Siskoudis, P., Marangos, M. and Sfikakis, P. 1988. *Metabolic effects of honey (alone or combined with other foods) in type II diabetics.* Acta Diabetol. Latina, 25(3): 197–203.

Kime, R.W., McLellan, M.R. and Lee, C.Y., 1991. *An improved method of mead production.* Am. Bee J., 131: 394–395.

Komosinska-Vassev, K., Olczyk, P., Kaźmierczak, J., Mencner, L. and Olczyk, K., 2015. *Bee pollen: chemical composition and therapeutic application.* Evid.-Based Complement. Altern. Med, 2015, 1–7.

Krell, R., 1996. *Value-added products from beekeeping.* FAO Agricultural Services Bulletin, 124, Food and Agriculture Organization of the United Nations, Rome, Italy.

Linskens, H.F. and Jorde, W. 1997. *Pollen as food and medicine—A review.* Econ. Bot., 51: 78–86.

Lücke, H. 1935. *Wundbehandlungmit Honig und Lebertran. (Wound treatment with honey and codliver oil).* Dtsch. Mediz. Wochenschrift, 41: 14–17.

Majno, G. 1975. *The healing hand: man and wound in the ancient world.* Harvard University Press, Cambridge, MA, 571 pp.

Mikhailov, A.C. 1950. *The application of medicated honey to eye diseases.* Pchelovodstvo, 2: 117–118.

Molan, P.C., Allen, K.L., Tan, S.T. and Wilkins, A.L. 1989. *Identification of components responsible for the antibacterial activity of Manuka and Viper's Bugloss honeys.* Ann. Conf. New Zealand Inst. of Chem., Paper No. 1.

Molan, P.C., Smith, I.M. and Reid, G.M. 1988. *A comparison of the antibacterial activity of some New Zealand honeys.* J. Apic. Res., 27(4): 252–256.

Morse, R.A. and Steinkraus, K.H. 1975. Wines from the Fermentation of Honey. In Honey: A Comprehensive Survey. E. Crane, ed.: 392–407.

Paillon, F. 1960. *La fabrication des produitsalimentaires au miel.* Girardot et Cie., Paris, 238 pp.

Panesar, P.S., Joshi, V.K., Bali, V. and Panesar, R. 2017. *Technology for Production of Fortified and Sparkling Fruit Wines.* Sci. Technol. Fruit Wine Product., 2017, 487–530.

Qureshi, N. and Tamhane, D.V. 1985. *Production of mead by immobilized whole cells of Saccharomyces cerevisiae.* Appl. Microbiol. Biotechnol., 21(5): 280–281.

Ramos, A.F.N. and Miranda, J.D. 2007. *Propolis: a review of its anti-inflammatory and healing actions.* J. Venom. Anim. Toxins Incl. Trop. Dis., 13: 697–710.

Revathy, V. and Banerji, S.A. 1980. *A preliminary study of antibacterial properties of Indian honey.* Ind. J. Biochem. Biophys. 17(Suppl. No. 242): 62.

Rodriguez Lopez, C. 1985. *Determinacin espectro-fotome'trica del color de las mieles.* Vida Apic., 16: 24–29.

Rombauer, I.S. and Rombauer Becker, M. 1975. *Joy of cooking.* Bobbs-Merrill Company, New York, NY, 915 pp.

Salem, S.N. 1981. *Honey regimen in gastrointestinal disorders.* Bull. Islamic Med., 1: 358–362.

Samarghandian, S. Farkhondeh, T. and Samini, F. 2017. *Honey and health: A review of recent clinical research.* Pharmacogn. Res., 9(2): 121.

Shambaugh, P., Worthington, V. and Herbert, J.H. 1990. *Differential effects of honey, sucrose, and fructose on blood sugar levels.* J. Manipul. Physiol. Therap., 13(6): 322–325.

Silva-Carvalho, R., Baltazar, F. and Almeida-Aguiar, C. 2015. *Propolis: a complex natural product with a plethora of biological activities that can be explored for drug development.* Evid. Based Complement. Altern. Med, 2015. 1–29, https://doi.org/10.1155/2015/206439

Subrahmanyam, V.M. 1993. *Storage of skin grafts in honey.* Bee Well 3(1): 6 (also in Lancet, 1/93)

Toth, G., Lemberkovics, E. and Kutasi-Szabo, J., 1987. *The volatile components of some Hungarian honeys and their antimicrobial effects.* Am. Bee J. (U.S.A.), 127 (7), 496–497.

Uccusic, P., 1983. *Doctor Bee: bee products, their curative power and application in medicine.* Ariston Verlag, Genf, Switzerland, 2nd edition in 1983, 198.

Wagh, V.D. 2013. *Propolis: a wonder bees product and its pharmacological potentials.* Adv. Pharm. Sci. 2013, 1–11, https://doi.org/10.1155/2013/308249

White, J.W. 1975a. Physical characteristics of honey. In: *Honey, a comprehensive survey,* Crane (ed.), Heinemann, London, UK, pp. 207–239.

White, J.W. 1975c. Honey. In: *The hive and the honeybee.* Dadant & Sons Inc., Hamilton, IL, pp. 491–530.

Zwaenepoel, C. 1984. Honey: facts and folklore. Alberta Beekeepers' Association, Edmonton, Canada, 24 pp.

5

Honeybees and Honey Enterprise Boon to Farmers and Unemployed Rural Youth for Socio-Economic Development

Aishwarya Sharma and Younis Ahmad Hajam
Career Point University, Hamirpur, India

CONTENTS

DOI: 10.1201/9781003175964-5

Introduction

The nursing of honeybees and their rearing to obtain various products as well as for pollination purpose is known as beekeeping. It can be regarded as an agro-business as it provides good income to farmers (Singh and Singh, 2006). In another way, an applied science of rearing honeybees for economic benefit of man is regarded as beekeeping. Beekeeping helps in improving economic figures as it contributes a lot by alleviating poverty by increasing household income as well as the economy at national and global levels. The production of honey throws light on the economic dynamics at global level (Pande et al., 2020). It also contributes to nutritional security, serves as a source of industrial raw material, gives emphasis on conservation of biodiversity and, thus, leads to enhancement of environmental resilience (Ajao and Oladimeji, 2013). People from almost every age-group can practice beekeeping such as men, women, children, retired persons etc. (Monga and Manocha, 2011). Some results in line with the findings of the Masuku and Mtshali (2012) and Moniruzzaman and Rahman (2009) stated that the majority of people from young age-group participated and it led to the creation of self-employment due to the successful promotion of beekeeping enterprise in rural areas. *Apis mellifera, Apis dorsata, Apis cerana, Apis florae* are the most common and known species of honeybees in the world as stated by Kebede and Adgaba (2011). *A. mellifera* is native of Europe and Africa while the rest others are natives of Asia. *A. mellifera* is somewhat aggressive in nature but is easy to handle if compared with other species of bees keeping in mind the honey yield. It also has an ability to easily adapt to the different climatic conditions and is highly productive (Masuku, 2013).

A sweet, viscous, sugary and sticky food substance produced by the honeybees is called honey. Bees make honey from the secretions of flora in the form of nectar and the processes involved include enzymatic activity, regurgitation and water evaporation. Honey wax is stored in structures known as honeycomb. Due to worldwide commercial production and consumption by humans, value of honey variety can be examined (Gupta, 2014). Honey constitutes readily accessible proteins, sugars, amino acids, sugars, vitamins, minerals in minimal amount and certain other pigments in microquantities, flavour, phenolic compounds, colloids and aroma. It provides about 46 calories in a one tablespoon (Hasam et al., 2020).

Production of honey has a great exporting potential in many industries like food, pharmaceutical and many more, but it still faces much problem in production as well as marketing, although problems can be solved with an efficient marketing system. In apiculture increasing colony numbers doesn't matters, what really matters is the efficiency level. Lack of information and knowledge leads to production which is not in accordance with the economic conventions; thus, it can be stated that production of honey falls behind at domestic markets in terms of quality (Gill, 1996). If adopted on scientific lines, it can create a substantial employment opportunity by providing additional income. Other bee products leading to good incomes are bee wax, royal jelly, bee pollen, venom, propolis. Bee wax is a secondary product obtained by beekeeping which provides authenticity. Bee wax is also used for making candles. The royal jelly is secreted by worker bees for feeding

FIGURE 5.1 Status of honey production and marketing in India.

queen and production of royal jelly can be expensive as well as labour-intensive. It is good for health and is capable of delivering optimum daily intake just in its spoon feed, whereas bee pollen are used as the dietary supplements and are easy to collect. Venom is a secretion that contains toxins such as apitoxin produced by bees, Propolis is a resinous mixture produced by honeybees after mixing saliva and bee wax. It is used as a sealant. Apart from this, honey has the highest share in beekeeper market as it is the main bee product. More the honey production, more is the income for honey producers. Apiculture is not just an additional income source, but it is also a sector that is supporting local communities and their development (Figure 5.1).

In 2020, the honey market has reached INR 19.2 billion in India. https://www.imarcgroup.com › Indian-honey-market.

It is believed that the number in terms of output will increase in coming future marking an enormous growth. Honey production is projecting and is on verge of gaining momentum in food and non-food applications in the country. Honey is also acting as a premium ingredient in nutraceuticals, thus acting as strong market drivers. Multifloral honey is the most popular honey type in India on the basis of flavour; however, other flavours like eucalyptus honey and cidar honey are also available. According to the seasonality reports, spring and autumn seasons are dominant in production level of honey followed by winter, summer and monsoon seasons. The largest channel of distribution is the business to consumer segment, although there are two types of channels: business to consumer and business to business. Before purchasing the products, issues like awareness and consciousness about health, nutrition and quality become essential distinctiveness factor for consumers (Magnusson et al., 2001).

Due to the presence of a large number of the manufactures, honey market in India is highly fragmented. A scheme known as "Development of beekeeping for improving crop yield" was launched by the government to organise awareness programs and training for beekeepers, so as to support honey processing plant establishments, and on this approach, market is projected to reach INR 28,057 million by 2024 in coming future (National Bee Board, 2021).

Honey as a Source of Employment

The production of honey has become a source of employment to many people across the country and world, due to its wide use and properties. Beekeeping is a well-accepted farming technology and is suited to the varied range of ecosystems of the region. About 1.5 lakh people other than beekeepers are getting employed in this sector. Women from rural areas are passionate about this type of activity of beekeeping and are able to successfully earning good income out of it.

Honey as a Food Supplement

Due to value-added properties of honey, it serves to be a great natural food supplement. Usage of honey serves as an important constituent in the diabetes management, plasma insulin level elevations and in dropping blood glucose levels in the patients. Considered as the wonderful gift by nature and produced by *A. mellifera*, the sweet sugary substance called honey is widely used as food (Andualem, 2013). There are different forms of honey such as comb honey which contains chunks of honey. Comb honey is recommended to beekeepers in the beginning as it is easy to produce, whereas its cost is not so high as compared to other types. The extracted honey is most preferred by most consumers. Products like whipped honey and honey butter are also obtained from the extracted honey, whereas chunk honey is a mixture of comb honey and extracted honey bottled together.

As far as the nutritional knowledge concerned and being a priority in the field of science, the understanding of types and assortment of foods including their nutritional value and composition as a human dietary life has been possible. By gaining knowledge, people can find link between type and combination of food to some extent and can, thus, find relation of food intake and health (Zheng et al., 2011).

Honey as a Medicine

Antibacterial properties of honey have proven to be helpful in healing wounds after many laboratory and clinical studies (Molan, 2002). Using the recipes of the traditional medicine with classic medical treatment along with the products of apiculture helps the diabetic patients in maintaining the normal insulin levels of blood and, thus, controlling the health condition (Bobiş et al., 2018). Honey also serves as a broad-spectrum antibiotic to which maximum bacteria are found to be sensitive. It was confirmed in a report by Alvarez-Suarez and co-workers that antimicrobial property resides in every type of honey. It is suggested that the formation of hydrogen peroxide plays an essential part in lowering the harmful effects of infections caused by bacteria (Cooper et al., 2002; Lusby et al., 2002).

An infection of eye which mainly occurs due to bacteria or microbiological agents causes itchiness and irritation in eye and in such condition honey because of its anti-bacterial property act as humectants and is proven to be a good treatment providing relief. Honey is also good for sore throat because it has antimicrobial action that kills bacteria and smoothens the throat. As an alternative therapy for infections, burns and skin ulcers, honey can be used as a natural remedy as it has antimicrobials that enhance immune system and causes reduction of inflammation and infection (Al-Waili et al., 2011) acidity of honey, its osmotic effect along with hydrogen peroxide and antioxidant content leads to immunity stimulation.

Honey as an Antioxidant

Honey being a therapeutic and a product of nutritional importance contains antioxidants. It is a natural substance which has many effects such antibacterial, hypoglycemic, hepatoprotective, reproductive, antihypertensive and antioxidative. Chronic diseases like diabetes mellitus, hypertension, cancer, atherosclerosis, Alzheimer disease are well known in today's world for their global prevalence. They have been the cause of mortality rates at global level, and the cause is absolutely linked to the term known as oxidative stress. Imbalance among the oxidants and antioxidants coming out in favour of oxidants, leading to damage is regarded as an oxidative stress. To overcome or reduce this problem of oxidative stress, antioxidants are essentially required (Pasupuleti et al., 2017).

Honey as a Raw Material in Many Industries

Honey-related industries have been acting like a giant leap and as an important medium acting as a source of food supplement for human beings. In international and national markets, honey acts as major consumable and is raw material for many industries like pharmaceutical, cosmetic, food and brewery (Yeow et al., 2013).

Famuyide et al. (2014) reported that various industries like pharmaceutical, food, cosmetic and even brewery are dependent on the honey production as it is the source for their raw material. According to the report by Agricultural and Processed Food Products Export Development Authority, India has exported 51547.31 metric ton of honey to world for about Rs 653.58 crores during 2017–2018 among which major export destinations were the United States, Saudi Arab, Canada and Qatar (Figure 5.2) (National Bee Board, 2021).

Honeybees as Pollinators

Honeybees reside in big perennial colonies and contribute a lot as pollinators. The production of genetically wide variety of the plants is possible only because of pollination and the fact is that about one-third of human food supply depends upon the

FIGURE 5.2   Honey processing plant.

pollination caused by insects (Dafni et al., 2005). Bee pollination helpful in agricultural sector as it causes about 50% increase in the production, especially *A. mellifera* which is an essential pollinator among the insects across the world, playing an important role in the preservation of natural resource (Klein et al., 2007).

It is indicated in the studies that beekeeping provides great benefits like employment, pollination of crops and biodiversity conservation. Generating income from hive products (Farinde et al., 2005) and giving bee colonies on rent for crop pollination (Masuku, 2013).

Worldwide Import and Export of Honey

Global honey production that enters into international trade accounts for only about 31.2% and the countries that are leading in honey export are China, Argentina, Mexico and India, whereas the countries which serve as major importers are United Kingdom, Japan, Germany and United States. Honey and other bee products have certain food safety and quality parameters which serve as the standard volume determinants for honey trade. Worldwide honey production is identified by its diversity, geographical as well as climatic factors. China constitutes about 35% honey production in the world followed by other countries like Turkey, Argentina, Ukraine, United States, Russia, India and Mexico which accounts for 6% as a total (Abrol, 2016).

Participation of Organisations

The governmental and non-governmental interventions that lead to the enhancement of the animal health to improve quantity and quality of the desired product by introducing modern technologies of beekeeping including equipment and training contributes to upliftment of apiculture. Traditional way of honey production is also encouraged and it serves as an employment opportunity for people of rural areas (Güngör, 2018).

Involvement of the Professional People

There is a need of more active participation by scientists, ethno-entomologists and apiculturists so that this sector can be more developed, and there can be improvement in the techniques and practices used by people. More awareness and development in this field will eventually lead to the great and handsome income from the global export (Schabel, 2006).

Status of honey production is regarded as male dominated in South West Nigeria. People here are although literate having moderate status in education. A number of hives and education are the essential factors which promote level of productivity. There is a very good opportunity for investors in this sector as it is found in the study that each naira spends more than 15 and as a profit half kobo is produced.

Beekeeping Scenario in Foreign Countries

In the United States, challenges like diseases in response to bee industry, interest of consumers act as a support to beekeepers at local as well as national level and as estimated economic value of beekeeping is of much importance as compared to the value of honey produced. It was observed in a congressional research service report that benefit of $15–$20 billion was made in terms of pollination. There are around 9600 registered beekeepers in Australia. Business which operates more than 500 hives accounts for about 62% of total honey production covering approx. 250 businesses. Moreover, it is also studied that businesses with 250 or less hives produce only 16% of Australian honey. The majority of income is produced from other businesses, sourced income, certain enterprises or government investments. Australian Bureau of Agricultural and Resource Economics (ABARE) was commissioned by the Australian Honey Bee Industry Council (AHBIC) and Rural Industries Research and Development Corporation (RIRDC) to conduct survey related to honeybee business to figure out the performance of related business so that we could reach to the solutions. The information obtained regarding the costs, investment, capital, finance, performance, age, education and socio-economic conditions of the industry and people associated with beekeeping will help in targeting future challenges in this sector (Rodriguez et al., 2003; Lunyamadzo, 2016). Pakistan is yet not able to export honey and honey products on large scale as it is not self-sufficient in this field. However, on the basis of environmental and other factors, a bee colony can produce about 36.28–54.43 kg of surplus or harvested honey whereas 4.53–8.16 kg of pollens in year on an average. Other bee products are also beneficial to country in terms of pollination. Value addition and production of by products may supplement the bee-keeper's income. In Pakistan, beekeeping is a profitable business and about 7000 beekeepers are rearing exotic species in modern beehives. About 300,000 colonies are producing about 10,000 tons of honey p.a. Presence of excellent bee flora present on wide area and favourable climate in the country provides great expansion of bees and various opportunities.

Ethiopia is considered as the great honey-producing country in Africa as well as in the world. As per records around 45,000 tons of honey production was seen in 2013 which is about 27% of African as well as 3% of world production hence as a result Ethiopia is regarded as the largest producer of honey in Africa and is ranked on 10th position in the world (Bruinsma, 2003).

In India, around 50 million hectare area is of bee dependent crops and if 3 bee colonies are kept per hectare, it would be beneficial in terms of pollination, but to meet this requirement about 150 million colonies are needed according to the estimated study. In India, we have 1.2 million bee colonies in active mode and to fulfil this target more than 100 million colonies are still required in response to more expansion of this field as beekeeping which would ultimately cause strengthening of agriculture sector in India. Agriculture Products Export Development Authority (APEDA) is a nodal agency that works in exports of honey under the ministry of Commerce and industry, Government of India. KVIC which stands for Khadi and village industries commission play an essential role in selling honey in some states like Gujarat. In Gujarat collections from wild bees providing the income for the rural people. Some NGOs and businessmen also take part in it; however, honey is being collected by the tribal community.

Punjab is among the leading state in beekeeping and almost 35,000 beekeepers which produce about 15,000 metric tonnes of honey which accounts for more than 39% of honey production in India. Punjab Agricultural University is the pioneer for introduction, establishment and multiplication of high yielding bee and eventually that has brought industrialisation and commercialisation in the country.

Problems Faced in Beekeeping

1. Excessive use of pesticides
2. Weather and climatic effect
3. Various bee diseases
4. Bee enemies like wasp etc.
5. Lack of technical knowledge and awareness
6. Wrong selection of bees
7. Deforestation leading to lack of adequate flora
8. Unavailability of infrastructure at various levels

Another factor that affects honey production in a country is gender as because of stings a lot of people find it tough to harvest and existing division of labour may result in limiting the women participation in beekeeping sector (Singh and Singh, 2006). Poor feeding during winter season has effect on the honey production. The colony leaves the area when it is not fed well and as a result yield of honey production is being affected. Sugar syrup is then given in feed for at least 6 weeks before the onset of very first flow of nectar and this initiates the production at time of foraging during nectar flow (Brodschneider and Crailsheim, 2010).

Socio-Economic Benefits

Traditional way of beekeeping is a diverse activity which leads to socio-economic concept and lesser the chances of dependence on only the conventional crop as an income source (Olarinde et. al., 2008). Many other sources of income can be generated by renting out hives to the farmers for pollination and, thus, selling of bees can be encouraged giving rise to the income. Bees are sold as a small nucleus known as 'nuc' and are transported and expanded to full size hives latern.

Small scale farmers having low capital investments are empowered by honeybee enterprise. (Farinde et al., 2005). Beekeeping also provides economic incentive to the local people for natural habitat preservation and for enhancement of environment status.

The national beekeeping influenced by the international market of honey with about 90% domestic production that is for export whereas producers and authorities took action which led to 50% of production at national level for consumption, and thus, the influence on beekeeping is increased by the consumer behaviour and is marked as an impact on economy (Pande and Azad Thakur, 2011). It is mentioned by the Bullock et al. (2008) that beekeeping is a great socio-economic activity in primary sector.

An activity which has great ecological and socio-economic importance is beekeeping and it is also regarded as the major foreign exchange earner of livestock activities as all the products obtained by beekeeping lead to generation of income as well as provides employment to many people in the rural areas (Mickels-Kokwe, 2006).

Production of honey provides social and economic benefits to various communities at rural areas and, thus, served to be attention seeker sector in eyes of farmers. Benefits can be economic i.e. in terms of income and benefits can even be social in terms of wide applications offered to the community. Beekeeping can be studied by the numerous facts emphasising on the economic impact (Bodescu et al., 2009; Ştefan et al., 2009).

Important Role in Agriculture and Ecosystem Services

Honeybees play an important role in agricultural sector and also enhance ecosystem services. There are many problems such as change in climate, invasive species and biodiversity loss which ultimately results in reduction of quality health as well as affect life span. Such problems lead to decrease in the profits and, thus, affect economy and income too. Apart from these factors predators, parasites and certain diseases are some other factors that affect this sector by affecting honey production in terms of quantity and quality. It is, thus, suggested that beekeepers would be able to control such damage by using less harmful pesticides in response to the vegetation growing near to project area or apiculture (Vercelli et al., 2021).

Sector Demands Awareness and Experience

Special trainings should be encouraged to improve the output quantity and quality which will then increase income and economy. Colony size should also be increased in beekeeping enterprise at basic level which will benefit smallholder farmers in improving their livelihood. The beekeepers who are much experienced can demonstrate and explain techniques and can give the trainings to other inexperienced apiarists, beginners and school students on apiculture under agribusiness training. There is an immense potential to provide employment to the rural and unemployed people in India, apiculture produces processing in the sector as the area consists of vast range of biodiversity which has large amount of flora for the bees. Small capital investment is enough to start this business. Raw material is not required in much amounts if compared with other industries although equipment like bee boxes, honey extractor, smoker, hive tools etc. are required to carry out the process. Honey production has been made easier by improved equipment like beehives (Figure 5.3) (Wakgari and Yigezu, 2021).

Institutional Support

The National Bee Board, Department of Agriculture and Cooperation, Delhi registers beekeepers, farmers, society, scientist and development workers, manufactures and traders involved in the honey processing development. It facilitated actions by making seed money available for the small farmers' Agri-business Consortium (SFAC) which has resulted the formation of National Bee Board as a registered society under societies Registration Act, XXI 1860 on 19 July 2000. (National Bee

FIGURE 5.3   People engaged in beekeeping.

Board, 2021) On May 2005, beekeeping is being included in NHM which stands for National Horticulture Mission. TRIFED which stands for Tribal Cooperative Marketing Development Federation, India is playing an essential role as it is providing training to the tribal people, generally about cultivation and harvesting of wild honey. KVIC Mumbai, PAU Ludhiana, IARI Delhi are R&D organisations working for diversification of the products by measuring bio-chemical and other properties of honey processing. Many international NGOs along with government are distributing honey buckets, bee colonies and hive tools to farmers in the villages for poverty alleviation (National Bee Board, 2021).

The honey price showed wide fluctuations because of country origin, types and factors, competition, direct indirect imports, quality and market segment for which it is addressed (Soares et al., 2017). High yields, quality of honey, disease resistance etc. are the advantages of the modern hives whereas there are some disadvantages like high price, pests, predators and absence of wax. Increasing technological demand of beekeeping requires enormous improvement of beekeeping skills and knowledge. The efficiency of production in beekeeping can be improved by introducing modern technical facilities and new innovations (Crescimanno et al., 2014).

Requirement of Labour

As more hives are added, the labour time should decrease as a result of experience. Although at the beginning it is expected that honey producers used to spend at least 28 hours every year for managing two hives including caring and harvesting. Installation of common processing units for all the beekeepers at village level with a fixed amount of processing depending upon the demand and supply. Processing and labelling then leads to marketing and income. Heavy investment and 24 hours working labour is not required in beekeeping; thus, it can even be integrated with other farming practice. Transportation is an another problem in apiculture as per kilometre, it will be costly in case services are given by companies; thus, honeybee houses associations are helpful in protecting the interest of apiculturists in support with this activity which reflects the need of labour and provide income to local participants (Abebe, 2009). The social subsistence in area depending upon the availability of valuable resources which leads to the employment and survival dependence is seldom used for production cost calculation (Popescu and Popescu, 2019).

Challenges Faced in Honey Production

It is important to have an evaluation of the production system so that the determinants for benefit and the main constraints of beekeeping sector can be identified especially the diseases related to bees and conventional agriculture (Western, 2020). There are many common diseases like nosema and viruses. Few examples are chalkbrood, European foulbroods, snotty brood etc. They lead to high mortality, to overcome such issue mite-resistant and disease resistant strains are being introduced (Fazlullah, 2017). For the destruction due to Varroa mite, VSH which stands for Varroa Sensitive Hygiene, a genetic line was introduced in 2006 which helped in reduction of colonies susceptibility to Varroa mites and resulted in stronger colonies with increasing populations of bees. Enhancement of yield, production practices and diseases in response to bees are the different biophysical aspects on which the apiculture research depends (Awraris et al., 2015; Gallmann and Thomas, 2012)

Ancient Evidence of Honey

Prehistoric period reveals about the evidence in association with hunting for honey and beekeeping as it is seen in Mesolithic rock paintings in countries like India, Africa and Spain and some illustrations were also figured out in Egypt between 2400 and 600 BC (Buchmann and Repplier, 2005; Roffet-Salque et al., 2015).

Natural Food for Bees Should be Encouraged

October to November and the spring season is known to be the best time for honey production, although honey production can be carried on whole year by encouraging the growth of different seasonal flora so that bees can get their food from the natural sources, to produce the best quality of organic honey. Thus, natural food sources for bees should be encouraged instead of the sugary artificial diet (Halter, 2011).

Sustainable Apiculture

Beekeeping will show improvement and will continuously grow preserving its existence in future and, thus, going to be regarding among the sustainable agricultural activity. A massive sector depends upon the natural environment covering producers, exporters and various companies leading to sustenance and profitability. Various parameters such as productivity, research, professional knowledge, protection and profitability should be properly checked (Magdoff et al., 2000). For the production of honey and other bee products, organic farming is possible with cooperation of beekeepers who carry out bee protection and biotechnology in an eco-friendly way. Switching to organic system is necessary to minimise the risk of bee extinction and to carry on migration beekeeping which has impact on the pollination and, thus, on the biodiversity. Influencing profitability of beekeeping enterprise includes ecological conditions such as quality of queen, quality of flora and management of the resources (Isah, 2018). Quality of food is an important aspect for the consumers before purchasing as people are conscious about health and nutrition (Magnusson et al., 2001). Beekeeping has a great impact on the pollination of commercial crops and, thus, is considered as an important element of food security in many countries. About 100 species of crops provide almost 90% of food supplies for 146 countries, out of which approximately 71 are pollinated by bees as per FAO estimation. Insect pollination is considered important for the natural ecosystem maintenance FAO estimation shows that from about nearly 100 species of crops, 71 are pollinated by bees, which provide about 90% of food supply for almost 146 countries (FAO, 2021). The activity of these insect pollinators is remarked as essential for the maintenance of ecosystems as its

FIGURE 5.4   Activity of insect pollinators help maintaining ecosystem.

use is linked with the reforestation projects and is even used as bioindicators (del Valle, 2020) (Figure 5.4).

Honey Marketing and Influence of Marketing Outlet

The beekeeper sells less to the consumer at final level in terms of internal honey marketing as a result the received price is generally dependent on the number and power market of those who are associated with the process. Honey is even used as an ingredient in production of certain food items like candies, cereals etc. and is used as a raw material in several industries like cosmetic etc. (Beckett, 2011). There has been research claiming that the price of organic honey is relatively high in the market due to its demand in the market (Jacks et al., 2011).

Farmers are capable in planning production status according to the current demand in the markets only because of improved information sources and marketing demands to make better decisions in response to the schedules about harvest, market and also to negotiate more in response to the traders. An essential farm household decision for selling the product at different marketing outlets reflects an effect on the household income, thus making outlet marketing an important choice (Tarekegn et al, 2017; Thakur et al, 2018).

It was pointed out by various empirical studies that choices made by smallholder farmers regarding the market outlet can be affected by various factors like endowments of resources, prices, cost of transportation and household characteristics as well (Berhanu et al., 2013). Advertising is a way to provide information about various products and this term is interconnected with pricing which also shows comparative information and of products and price sensitivity. Product differentiation and uniqueness which is uniqueness is a measure for brand equity created as a result of advertising (Thakur et al., 2017). For the sake of health, honey vendors sell bee products. People are conscious about preservation and health improvement of their family (Schifferstein and Ophuis, 1998). For improving marketing so that it could be helpful for the poor people, implementation of market outlets and different trades at national and international level is essential so that farmers can choose their alternative outlet for selling their supply to increase the profit which eventually may increase income in association with household concept in response to the apiculture sector promoting creation of new design, innovation and policy (Chahal and Bala, 2010; Thakur et al., 2020).

Beekeeping – Creating and Strengthening Sustainable Livelihood for Rural and Poor People

Beekeeping has been proven to be a useful means of strengthening livelihood for people at rural areas as it forms and create various assets emphasising on physical, social and economic aspects. Apart from honey, beekeeping also produces many other products like pollen, propolis, royal jelly etc., thus, provides various opportunities and good income sources for people imparting the essence of value-added products, cultural values and respect. Experiences, capabilities as well as initiatives are very important terms as these are used in combining assets and beneficial in getting good incomes so as to strengthen livelihoods of the rural families depending upon the production strategies (Losch et al., 2012). Beekeeping has proven to be an essential branch of agriculture which is involved in its contribution to high-value products for people in terms of food and health; moreover, the sector is linked to biodiversity conservation and pollination as well. Benefits and employment generation ways from beekeeping includes processing and marketing of bee products such as honey, bee wax, propolis, pollens, royal jelly and bee venom as bee products. Further, multiplication and selling of bees, queen bees and nucleus colonies, reared queen bees having good economic traits such as resistance against disease and high yield production can lead to good price. Moreover, manufacturing and selling beekeeping equipment also provides employment opportunity (Guiso et al., 2004, 2009; Tarekegn et al., 2017).

REFERENCES

Abebe, A., 2009. *Market chain analysis of honey production in Atsbi Wemberta District, eastern zone of Tigray national regional state* (Doctoral dissertation, Haramaya University).

Abrol, D.P., 2016. Current scenario of beekeeping in India-challenges and opportunities. *Beneficial Insect Farming-Benefits and Livelihood Generation (ICAR-Indian Institute of Natural Resins and Gums, Ranchi, India, 2016)*, p. 11.

Ajao, A.M. and Oladimeji, Y.U., 2013. Assessment of contribution of apicultural practices to household income and poverty alleviation in Kwara State, Nigeria. *International Journal of Science and Nature*, 4(4), pp. 687–698.

Al-Waili, N., Salom, K. and Al-Ghamdi, A.A., 2011. Honey for wound healing, ulcers, and burns; data supporting its use in clinical practice. *The Scientific World Journal*, 11, pp. 766–787.

Andualem, B., 2013. Combined antibacterial activity of stingless bee (*Apis mellipodae*) honey and garlic (*Allium sativum*) extracts against standard and clinical pathogenic bacteria. *Asian Pacific Journal of Tropical Biomedicine, 3*(9), pp. 725–731.

Awraris, G., Amenay, A., Hailemariam, G., Nuru, A., Dejen, A., Zerihun, T. and Asrat, T., 2015. Comparative analysis of colony performance and profit from different beehive types in southwest Ethiopia. *Global Journal of Animal Scientific Research, 3*(1), pp. 178–185.

Beckett, S.T. ed., 2011. *Industrial chocolate manufacture and use.* John Wiley & Sons Ltd, West Sussex, United Kingdom.

Berhanu, B., Melesse, A.M. and Seleshi, Y., 2013. GIS-based hydrological zones and soil geo-database of Ethiopia. *Catena, 104*, pp. 21–31.

Bobiş, O., Dezmirean, D.S. and Moise, A.R., 2018. Honey and diabetes: the importance of natural simple sugars in diet for preventing and treating different type of diabetes. *Oxidative Medicine and Cellular Longevity, 2018*, p. 12. https://doi.org/10.1155/2018/4757893

Bodescu, D., Stefan, G. and Paveliuc, C.O., 2009. The comparative profitability of Romanian apiarian exploitations on size categories. *Bulletin of University of Agricultural Sciences and Veterinary Medicine Cluj-Napoca. Horticulture, 66*(2), p. 514.

Brodschneider, R. and Crailsheim, K., 2010. Nutrition and health in honey bees. *Apidologie, 41*(3), pp. 278–294.

Bruinsma, J. ed., 2003. *World agriculture: towards 2015/2030: an FAO perspective.* Earthscan Publications Ltd, London, United Kingdom.

Buchmann, S. and Repplier, B., 2005. *Letters from the hive: An intimate history of bees, honey, and humankind.* Bantam Sciences, New York City.

Bullock, C., Kretsch, C. and Candon, E., 2008. *The economic and social aspects of biodiversity: benefits and costs of biodiversity in Ireland.* Stationery Office, Government of Ireland.

Chahal, H. and Bala, M., 2010. Confirmatory study on brand equity and brand loyalty: a special look at the impact of attitudinal and behavioural loyalty. *Vision, 14*(1–2), pp. 1–12.

Cooper, R.A., Halas, E. and Molan, P.C., 2002. The efficacy of honey in inhibiting strains of *Pseudomonas aeruginosa* from infected burns. *The Journal of Burn Care & Rehabilitation, 23*(6), pp. 366–370.

Crescimanno, M., Galati, A. and Bal, T., 2014. The role of the economic crisis on the competitiveness of the agri-food sector in the main Mediterranean countries. *Agricultural Economics, 60*(2), pp. 49–64.

Dafni, A., Kevan, P.G. and Husband, B.C., 2005. Practical pollination biology. *Practical pollination biology.* Enviroquest, Ltd.

del Valle, E.R., 2020. *How Does Landscape Restoration in La Junquera Farm Affect Pollination & Biological Control Services by Insects on Almond Cultivation.* Wagningen, University and Research.

Famuyide, O.O., Adebayo, O., Owese, T., Azeez, F.A., Arabomen, O., Olugbire, O.O. and Ojo, D., 2014. Economic contributions of honey production as a means of livelihood strategy in Oyo State. *International Journal of Science and Technology, 3*(1), pp. 7–11.

FAO 2021. http://www.fao.org/3/x0083e/x0083e00.htm

Farinde, A.J., Soyebo, K.O. and Oyedokun, M.O., 2005. Improving farmers attitude towards natural resources management in a democratic and deregulated economy: Honey production experience in Oyo state of Nigeria. *Journal of Human Ecology, 18*(1), pp. 31–37.

Fazlullah, M., 2017. *Study on honey production by using wooden and poly hive in different seasons in Bangladesh* (Doctoral dissertation, Department Of Entomology, Sher-E-Bangla Agricultural University, Dhaka).

Gallmann, P. and Thomas, H., 2012. Beekeeping and honey production in southwestern Ethiopia. *Ethiopia Honey Bee Invest*, 10(3), pp. 99–109.

Gill, R.A., 1996. The benefits to the beekeeping industry and society from secure access to public lands and their melliferous resources. *Report to the Honeybee Research and Development Council of Australia.*

Guiso, L., Sapienza, P. and Zingales, L., 2004. The role of social capital in financial development. *American Economic Review*, 94(3), pp. 526–556.

Guiso, L., Sapienza, P. and Zingales, L., 2009. Cultural biases in economic exchange?. *The Quarterly Journal of Economics*, 124(3), pp. 1095–1131.

Güngör, E., 2018. Determination of optimum management strategy for honey production forest lands using A'WOT and conjoint analysis: a case study in Turkey. *Applied Ecology and Environmental Research*, 16(3), pp. 3437–3459.

Gupta, R.K., 2014. Technological innovations and emerging issues in beekeeping. In *Beekeeping for poverty alleviation and livelihood security* (pp. 507–554). Springer, Dordrecht.

Halter, R., 2011. *The Incomparable Honeybee & the Economics of Pollination: Revised & Updated*. Rocky Mountain Books Ltd., Ingram Publisher Services (USA).

Hasam, S., Qarizada, D. and Azizi, M., 2020. A review: honey and its nutritional composition. *Asian Journal of Research in Biochemistry*, 7, pp. 34–43.

Isah, I.A., 2018. Profitability of stingless beekeeping production and beekeepers' perception of extension agents' performance. Ministry for Primary Industries, Wellington, New Zealand. http://www.mpi.govt.nz/

Jacks, D.S., O'rourke, K.H. and Williamson, J.G., 2011. Commodity price volatility and world market integration since 1700. *Review of Economics and Statistics*, 93(3), pp. 800–813.

Kebede, A. and Adgaba, N., 2011. *Honey Bee Production Practices and Honey Quality In Silti Wereda Ethiopia* (Doctoral dissertation, Haramaya University).

Klein, A.M., Vaissiere, B.E., Cane, J.H., Steffan-Dewenter, I., Cunningham, S.A., Kremen, C. and Tscharntke, T., 2007. Importance of pollinators in changing landscapes for world crops. *Proceedings of the Royal Society B: Biological Sciences*, 274(1608), pp. 303–313.

Losch, B., Fréguin-Gresh, S. and White, E.T., 2012. *Structural transformation and rural change revisited: challenges for late developing countries in a globalizing world.* World Bank Publications.

Lunyamadzo, M.G., 2016. *Performance and contribution of beekeeping enterprises to livelihood in Songea District* (Doctoral dissertation, Sokoine University of Agriculture).

Lusby, P.E., Coombes, A. and Wilkinson, J.M., 2002. Honey: a potent agent for wound healing?. *Journal of WOCN*, 29(6), pp. 295–300.

Magdoff, F., Foster, J.B. and Buttel, F.H. eds., 2000. *Hungry for profit: The agribusiness threat to farmers, food, and the environment*. NYU Press, New York, NY.

Magnusson, M.K., Arvola, A., Hursti, U.K.K., Åberg, L. and Sjödén, P.O., 2001. Attitudes towards organic foods among Swedish consumers. *British Food Journal*, 103 (3), pp. 209–227.

Masuku, B.S. and Mtshali, M.N., 2012. Challenges faced by apiculture SMMES in Northern Hhohho, Swaziland. *Loyola Journal of Social Sciences*, 26(1), pp. 141–165.

Masuku, M.B., 2013. Socioeconomic analysis of beekeeping in Swaziland: a case study of the Manzini Region, Swaziland. *Journal of Development and Agricultural Economics*, 5(6), pp. 236–241.

Mickels-Kokwe, G., 2006. *Small-scale woodland-based enterprises with outstanding economic potential: the case of honey in Zambia*. Cifor. National Library of Indonesia Cataloging-in-Publication Data.

Molan, P.C., 2002. Re-introducing honey in the management of wounds and ulcers-theory and practice. Ostomy/Wound Management *48*(11), pp. 28–40.

Monga, K. and Manocha, A., 2011. Adoption and constraints of beekeeping in District Panchkula (Haryana), India. *Education*, *10*(16.6), p. 3.

Moniruzzaman, M. and Rahman, M.S., 2009. Prospects of beekeeping in Bangladesh. *Journal of the Bangladesh Agricultural University*, 7(1), pp. 109–116.

National Bee Board, 2021, https://nbb.gov.in/index.html; http://nhb.gov.in/beekeeping_report.aspx?enc=3ZOO8K5CzcdC/Yq6HcdIxMDNqXAfCKV7Vr4L5zsSZ1A

Olarinde, L.O., Ajao, A.O. and Okunola, S.O., 2008. Determinants of technical efficiency in bee-keeping farms in Oyo State, Nigeria: a stochastic production frontier approach. *Research Journal of Agriculture and Biological Sciences*, *4*(1), pp. 65–69.

Pande, R., Ramkrushna, G.I., Verma, P. and Shah, V., 2020. Role of honey bees for income generation in farming system. *Biotica Research Today*, *2*(11), pp. 1122–1125.

Pande, R., Azad Thakur, N.S., 2011. Promotion of scientific beekeeping in Meghalaya. ICAR RC for NEH region Umiam-793 103 Meghalaya. Technical Bulletin No. 72.

Pasupuleti, V.R., Sammugam, L., Ramesh, N. and Gan, S.H., 2017. Honey, propolis, and royal jelly: a comprehensive review of their biological actions and health benefits. *Oxidative Medicine and Cellular Longevity*, 2017. 21 pp. https://doi.org/10.1155/2017/1259510.

Popescu, C.R.G. and Popescu, G.N., 2019. The social, economic, and environmental impact of ecological beekeeping in Romania. In *Agrifood economics and sustainable development in contemporary society* (pp. 75–96). IGI Global.

Rodriguez, V.B., Riley, C., Shafron, W. and Lindsay, R., 2003. Honeybee industry survey. *Rural Industries Research and Development Corporation, Pub (03/039)*. Centre for International Economics Canberra & Sydney, http://www.thecie.com.au/

Roffet-Salque, M., Regert, M., Evershed, R.P., Outram, A.K., Cramp, L.J., Decavallas, O., Dunne, J., Gerbault, P., Mileto, S., Mirabaud, S. and Pääkkönen, M., 2015. Widespread exploitation of the honeybee by early Neolithic farmers. *Nature*, *527*(7577), pp. 226–230.

Schabel, H.G., 2006. *Forest entomology in East Africa: forest insects of Tanzania*. Springer Science & Business Media.

Schifferstein, H.N. and Ophuis, P.A.O., 1998. Health-related determinants of organic food consumption in the Netherlands. *Food Quality and Preference*, *9*(3), pp. 119–133.

Singh, D., Singh, D.P., 2006. *A handbook of beekeeping*. Agrobios, India, 287.

Soares, S., Amaral, J.S., Oliveira, M.B.P. and Mafra, I., 2017. A comprehensive review on the main honey authentication issues: production and origin. *Comprehensive Reviews in Food Science and Food Safety*, *16*(5), pp. 1072–1100.

Ştefan, G., Bodescu, D. and Boghiţă, E., 2009. Analysis of the economic size of beekeeping holdings in Romania. *Lucrări Ştiinţifice, Universitatea de Ştiinţe Agricole Şi Medicină Veterinară "Ion Ionescu de la Brad" Iaşi, Seria Horticultură*, *52*, pp. 361–364.

Tarekegn, K., Haji, J. and Tegegne, B., 2017. Determinants of honey producer market outlet choice in Chena District, southern Ethiopia: a multivariate probit regression analysis. *Agricultural and Food Economics*, *5*(1), pp. 1–14.

Thakur, P., Mehta, P. and Gaurav, K.K., 2018. Farmers' perceptions towards applications of modern farming tools used in vegetable production at Solan district, Himachal Pradesh-India. *Indian Journal of Pure & Applied Biosciences SPI*, 6(3), pp. 617–623.

Thakur, P., Mehta, P. and Gupta, N., 2017. An impact study of food product packaging on consumer buying behaviour: a study premise to Himachal Pradesh-India. *International Journal of Bio-resource and Stress Management*, 8(6), pp. 882–886.

Thakur, P., Mehta, P., Dhiman, R. and Kumar, K., 2020. An assessment of awareness level and modern farm business management practices adopted by vegetable farmers' in Mid-hills of Himachal Pradesh. *Indian Journal of Extension Education*, 56(2), pp. 143–148.

Vercelli, M., Novelli, S., Ferrazzi, P., Lentini, G. and Ferracini, C., 2021. A qualitative analysis of beekeepers' perceptions and farm management adaptations to the impact of climate change on honey bees. *Insects*, 12(3), p. 228.

Wakgari, M. and Yigezu, G., 2021. Honeybee keeping constraints and future prospects. *Cogent Food & Agriculture*, 7(1), p. 1872192.

Western, B.B., 2020. *Investigating the need for and interest in offering honey bee pollination services in apple production: perspectives from growers and beekeepers in Midt-Telemark, Norway* (Master's thesis, Norwegian University of Life Sciences, Ås).

Yeow, S.H.C., Chin, S.T.S., Yeow, J.A. and Tan, K.S., 2013. Consumer purchase intentions and honey related products. *Journal of Marketing Research & Case Studies*, 2013, p. 1.

Zheng, H. Zheng, H.Q., Wei, W.T. and Hu, F.L., 2011. Beekeeping industry in China. *Bee World*, 88(2), pp. 41–44.

6

Composition of Honey, Its Therapeutic Properties and Role in Cosmetic Industry

Indu Kumari
Arni University, Kangra, India

Ankush Sharma
Sri Sai University, Kangra, India

Diksha, Preeti Sharma and Younis Ahmad Hajam
Career Point University, Hamirpur, India

Rajesh Kumar
Himachal Pradesh University, Shimla, India

CONTENTS

DOI: 10.1201/9781003175964-6

Introduction

Honey is one of the most valued, natural substances known to humankind since the early period. Honey consumption as a diet and medicine has high nutritional and therapeutic value. As a natural sweet substance, it is processed and produced by honeybees from the nectar of flowers. Honey is a natural product produced by the honeybees and is not altered or blended with other substances. The properties of honey, such as antioxidant, antibacterial, anti-fungicidal, hepatoprotective, and anti-inflammatory, are widely known (Ajibola, 2015).

Composition of Honey

Physical Characteristics of Honey

Physical characteristics of honey include viscosity, crystallisation, hygroscopicity, electrical conductivity, optical properties, surface tension and colour (Ball, 2007). Mechanism of separation, refining, rapping and maintenance are the causes of physiognomy in different honey samples. The physical properties of honey are discussed in the following sections.

Viscosity

Viscosity is defined as fluid's resistance to flow. Viscosity is chiefly dependent upon the honey's moisture content, temperature and chemical constituents (Ajibola, 2015; Nayik *et al.*, 2018). Honey having higher moisture content flows faster than that having lowered one, and it is very important for the shelflife of honey. This is due to an increase in honey fermentation by certain osmotolerant yeast (Abdulkhaliq and Swaileh, 2017). Thus, thickness plays a significant role in the quality specification in different manufacturing phases of honey-like honey extraction up to distortion, transportation, preparation and wrapping. Therefore, it is important in different applications like handling honey, maintenance, mechanism, shifting, process assurance and receptive evaluation of food (Nayik *et al.*, 2018). According to Codex Alimentations

Standard, moisture content of honey should be <20%. The foaming characteristics of some honey are also due to the viscosity and surface tension (Machado De-Melo *et al.*, 2017). Higher viscous honey provides hindrance to homogenisation in preparation and regulates the utilisation in Food Company and promoting 'dry honey' especially in confectionery beverages (Tong *et al.*, 2010).

Hygroscopicity

Hygroscopicity is the property in which honey can soak up the humidity present in the atmosphere. It may increase the moisture content in the honey that leads to fermentation. The moisture content absorbed by honey depends upon the relative humidity present in the atmosphere. The hygroscopicity of honey is characterised by a large amount of sugar present in it. Among all sugar present in the honey, fructose is the most hygroscopic. Therefore, hygroscopicity increases the H_2O constituents present on the upper layer that change the aspect of honey at the time of stocking and can affect its quality during storage (Bartulović, 2015).

Optical Rotation (Polarisation)

Optical rotation means oriental pathway of the circulation about the optical axis by liquid substances. Honey shows the property of rotating the plane-polarised light due to its composition of carbohydrate present in it. Each sugar has its angle-specific optical rotation (specific rotation). So, this property of honey is mainly due to fructose and glucose. Fructose can turn the 'polarised light' angle opposite to right, which leads to the −ve 'optical rotation value (fructose $[\alpha]D20 = -92.4°)$', or other towards the right which shows +ve 'optical activity (glucose$[\alpha]D20 = +52.7°)$' (García-Alvarez *et al.*, 2002; Sanz *et al.*, 2002; Dinkov, 2003). Therefore, it is the congregation of the distinct sweeteners in the honey responsible for the overall value of the optical rotation (Bogdanov *et al.*, 1999). Floral honey has a great number of fructose constituents which are left-handed. Otherwise, honeydew honey or adulterated floral honeys are dextrorotatory. Honeydew honey usually contains the lower fructose content and contain melezitose and lose showing specific optical rotation ($[\alpha]$ D20 = +88.2°) and ($[\alpha]$ D20 = +121.8°), respectively. Hence, this property can be used to distinguish between nectar and honeydew honey or adulterated honey. In some countries (Greece, Italy and United Kingdom), the quantification of particular spinning has been utilised because it has a common specific rotation, i.e., positive (Piazza and Persano Oddo, 2004). This property may be due to compositional differences of honey due to different botanical origins (Bertoncelj *et al.*, 2011). Hence it may be used to determine the adulteration (Machado De-Melo *et al.*, 2017).

Refractive Index

It is perceptible affection which extends within '1.504 & 1.4815'and increases according to the climate condition while having a large amount of hard particle and a little bit of H_2O (Sinz-Laın and Gomez-Ferreras, 2000). The refractive index can measure honey's moisture content using the Wedmore Table (Nikolova *et al.*, 2012).

Density and Specific Gravity

Specific gravity is also known as relative density. This property of honey is greatly affected by huge saccharide (sugar) substances. Density can be indicated as gravitational attraction, and the value depends upon the H_2O, climate and hard material quantification. Solidity reduces directly with an increase in climate or moisture substance and expands directly through a rise in the hard matter (Sabatini, 2007; Oroian, 2013). However, the average specific gravity of honey at 200 Celsius ranges up to '1.40 to 1.44 g/ml', corresponding to the biological source of honey (Crane, 1980; Sainz-Laın and Gomez-Ferreras, 2000). During storage in tanks, honey having density variation shows the stratification having a first surface containing minimum solidity and large humidity content, making it extra susceptible to volatilisation (White, 1975; Krell, 1996; Bogdanov, 2011b). Therefore, appropriate processing technology should be adopted along with its packaging standard. Relative density also correlates with the climatic condition of a region due to humidity present in the atmosphere. This is true to the region where humidity decreases from the north to the south.

Granulation and Crystallisation

Honey is a supersaturated solution of a variety of sugars. Among the sugars, glucose content in the honey is more liable to crystallisation. Glucose having less solubility than $C_6H_{12}O_6$ sets apart out of H_2O and gives rise to a concentrated mixture and forms the dextrose micronised liquid. H_2O that becomes free from the glues increases the water content of the uncrystallised honey and makes it prone to fermentation and spoilage. However, the fructose stayed in liquid form and gathered in the thin layer around the glucose crystals. Honey changes its colour and transforms into brighter colour and changes its translucent nature and changes its taste. Honey containing higher fructose content, i.e., 'acacia and sage', tending to stay in a fluid state for an extended time, and honey containing a high concentration of glucose disintegrated rapidly next collecting or in the cell such as rape or dandelion honey (Maurizio, 1962). The honeydew honey's glucose and melibiose content increases the honey's crystallisation process if they are more than 28–30% in floral honey and more than 10% in honeydew honey, respectively. Crystallisation of honey is affected by various temperature ranges, and absolute condition forms within '10°C to 18°C and 14°C' are considered comfortable febricity. Crystal formation becomes decreased at lower temperature range, although lower temperature decreases glucose solubility enhancing the granulation. This is due to an increase in honey viscosity which makes the diffusion of glucose crystal more difficult to move. However, the temperature range lies between 5 and 7°C, facilitating the formation of crystallisation nuclei. The storage of honey at low febricity shows 'Newtonian, pseudoplastic or thixotropic' features, subjected to its size, amount and crystal arrangement.

Crystal

Crystal formation of honey samples containing higher moisture contents is favoured at −20°C in contrary to honey at ambient conditions (Conforti *et al.*, 2006). Concerning moisture content, crystallisation of honey favours optimally within 15 and 18% of humidity content (Bogdanov, 2011a). Moisture generally reduces the process of formation of a crystal due to the reduction of glucose congestion. If the dextrose gets precipitated, leaving behind the fructose on the surface of the liquid, sugar absorbs H_2O from the atmosphere and makes it prone to the risk of fermentation (Machado De-Melo *et al.*, 2017). Normally, crystallisation of sugar (honey) occurs when kept in lousy and independent mode; it makes outcomes look dark and hence, wrongly perceived as unadulterated honey to the consumer. Contrary to this, cream crystallised honey is a desirable product for the commercialisation of spread honey. It contains very small-sized crystals in large numbers not to be perceived by the palate (Conforti *et al.*, 2006; Machado De-Melo *et al.*, 2017).

Colour

Honey possesses a wide range of colour variations from clear, colourless, yellow, amber or black, or maybe 'green or reddish reflexes'. The colour of honey is associated with botanical origin, climatic conditions, duration and storage conditions. Some colours are specific to some particular honey: 'bright yellow of sunflower, reddish undertones of chestnut, greyish of eucalyptus and greenish of honeydew'. Several substances influence the colour of honey, namely, various products like glucose, fructose, sucrose, etc.; tetraterpenoids, phylloxanthins, 'anthocyanins, amino acids and phenolic compounds, mainly flavonoids' and mineral content are present. Generally, dark honey contains higher mineral content, polyphenol, high acidity and dextrin. Dark honey consists of melanoidin that is formed from the carbonyl reaction (Maillard reaction) that takes place between 'amino acids and sugars in an acid medium'; the caramelisation process of sugars due to acidity of honey, increase in tincture material of honey through oxidisation of macromolecule, can form brown complexes with amino acid and protein and lipid oxidation reactions leading to the formation of carbonic compound if kept for longer duration at freezing condition. Heating is also responsible for the change in the countenance of honey, i.e., shade, formation of crystal, flavour and aroma. Indeed sugar (natural honey) also changes to dusk after boiling (Ajibola, 2015).

pH

Generally, all honey samples show acidic pH (pH < 4.5), a typical characteristic of blossom honey. However, the standard limit for the pH values ranges from 3.40 to 6.10 and ensures the freshness of honey samples. The high acidity of honey corresponds to organic acid produced by the fermentation of sugars present in the honey. It is accountable for two important characters, i.e., flavour and stability of honey (Sohaimy *et al.*, 2015). The variation in the pH range of honey observed in different honey samples is mainly due to the differences in acid and mineral content of different honey samples. The pH values have an impact on the texture, stability and shelf life of honey. Hence, this parameter is of great significance during the extraction process of honey and storage (Gomes *et al.*, 2009).

Osmotic Pressure

The chemical composition of sugar is 85–95%, in which glucose, fructose, sucrose and maltose range between 28–31, 22–38, 1–4 and 1–9%, respectively. The high osmotic pressure of honey having a low potential of hydrogen, acetic acid and H_2O_2 formed by enzyme activity on glucose inhibits the microorganism development, consequently enhancing the life period of the substance (Manyi-Loh *et al.*, 2011; Alvarez-Suarez *et al.*, 2014; Silva *et al.*, 2017). However, it is suggested by a researcher that the ability of honey to fight infection comes from two separate mechanisms: (1) bacterial effect through an unknown source which is dependent on heavenly food and (2) other is consequence on safety 'QS systems' of bacteria which are related to the glucose material present and not depend on the origin of nectar.

Chemical Characteristics of Honey

There are wide variations in the composition of honey reflected in colour, flavour and taste. It varies from floral source to its origin. Natural honey is a complex nutritional food substance primarily comprised of (60–85%) carbohydrates and (12–23%) water. In addition, honey also contains organic acids, minerals, vitamins, enzymes, proteins, amino acids, Maillard reaction products, charged constituents in small quantities, and more than 300 plant derivative and also having pollen grains (Almeida-Muradian *et al.*, 2013; Ajibola, 2015). The chemical properties of honey are described in the following sections.

Moisture Content

Water is also a significant component and also an important quality parameter of honey. The moisture content of honey may vary from 15 to 23%. Normally, honey extracted from well-sealed combs contained a water content of less than 18%. According to the Codex and European Directive, moisture content of honey should be less than 20%, except for (Calluna vulgaris honey) heather honey that is permitted up to 23%. The moisture content of honey is related to climatic conditions, the variety of the bees, the bee colony strength, the humidity and the air temperature in the hive, the preparation and stored environment and the botanical origin of honey. It also affects some physical characteristics of honey, i.e., viscosity, specific weight, maturity, flavour, crystallisation and specific gravity (do Nascimento *et al.*, 2015). It is an important factor that influences the shelf life of honey. Honey having higher moisture content may prone to evaporation due to OP (osmotic pressure) of sugars is not much powerful to inhibit the proliferation of osmophilic (sugar-tolerant) yeast. Therefore, the chances of fermentation increase with an increase in water content. However, some researchers took into account the water activity (Aw) that is responsible for the growth of microorganisms. The reaction of H_2O is the quantity of free H_2O presence for the growth of microorganisms. The water activity of honey varies from 0.49 to 0.65; it can extend up to 0.75 for some honey (Cavia *et al.*, 2004; Costa *et al.*, 2013). The requirement of H_2O microbial actions growth is up to '0.90 for bacteria, 0.80 for yeast and 0.70 for moulds'. Aw values less than 0.60 prevent the development of symbiotic bacteria (yeast), which results in honey volatilisation (Bogdanov, 2011b).

Sugars Content

Honey includes H_2O and sugars that are accountable for 95–99% of honey dry matter and 4–5% of plant sugar (Ajibola, 2015). Generally, fructose and glucose in honey are (32–44%) and glucose (23–38%), respectively. However, the proportion can vary in different honeys. Besides these, galactose and lactose also have been reported from the same honey samples in a small amount (Val *et al.*, 1998). Fructose is the predominant sugar in most honey samples, i.e., Acacia and chestnut honey, but *Brassica napus*, dandelion (*Taraxacum officinale*) and blue curls (*Trichostema lanceolatum*) kinds of honey having the glucose as main sugar. Besides these, galactose and lactose also have been reported from some honey samples in a small amount (Val *et al.*, 1998). Besides this, various sugars disaccharides, trisaccharides and other oligos- and polysaccharides present in comprised about 5–15% of the honey. These include like 'maltose, sucrose, turanose, trehalose, gentiobiose, isomaltose, kojibiose, raffinose, erlose, melezitose, maltotriose, panose, isomaltotriose and maltotetraose' (Machado De-Melo *et al.*, 2017). Sucrose and maltose are the most common disaccharides present in honey. High sucrose may be related to honey adulteration. However, this may be due to the immaturity of honey, botanical origin and high nectar flux. Sugars present in the honey are produced by the bee's transformation of sucrose present in the nectar by the activity of invertase enzymes from the salivary glands of the bees. The invertase enzyme also possesses transglycosylation activity giving rise to complex sugars from monosaccharides (Machado De-Melo *et al.*, 2017). There are set criteria for whole monosaccharide material that should be more than 60 g/100 g for nectar honey, and for honeydew, it is more than 45 g/100 g. Sucrose content should be less than 5 g/100 g except for some specific honey (Council Directive 2001/110/ EC). The fructose: glucose ratio in honeydew honey is higher than nectar honey (Jakubik *et al.*, 2020).

Acid

Honey contains various organic acids found associated with their corresponding gluconolactone in various equilibriums. They contribute about <0.5% of whole solid in honey still responsible for flavour, fragrance, shade or maintenance of honey, for reducing the development of microbe. Predominant natural acid present in honey, i.e., $C_6H_{12}O_7$ represents 70–90% of the total and is induced through oxidoreductase activity on $C_6H_{12}O_6$. Other organic acid includes CH_3COOH, $C_3H_4O_3$ and $C_4H_6O_6$ (Mato *et al.*, 2003; Ahmed *et al.*, 2018). Out of these, some are intermediate in the Krebs cycle and other enzymatic pathways (Machado De-Melo *et al.*, 2017). Generally, honeys having acidity are dark in colour and storage of honey for longer increases acidity. A low concentration of organic acid characterises some kinds of honey, therefore, lighter in colour, i.e., Acacia, chestnut and meadow. Some organic acid maybe act as the marker for distinguishing particular honey samples. The concentration of $C_6H_8O_7$ can be utilised as good variability for the polarity of two important kinds of honey, such as 'floral and honeydew honey'. Generally, honeydew honey contains a higher concentration of citric acid, i.e., oak and pine honeydew and thymus honey (del Nozal *et al.*, 1998; Haroun *et al.*, 2012).

Proteins and Amino Acid

Honey contains protein in a small amount. Honey protein originates from the 'salivary glands of bees and nectar', pollens or honeydew of plants. Generally, honey contains a very little amount of protein ranges from 0.05 to 0.1%, whereas in 'honeydew honey', this amount is 3.0% (Czipa *et al.*, 2012; da Silva *et al.*, 2016). However, a significantly higher concentration of proteins has been found in litchi honey of Bangladesh and honey from Sundarban of Bangladesh (Rahman *et al.*, 2017). The presence of proteins in the honey contributes to having a lower surface tension otherwise, showing a marked tendency of foaming and air bubbles retentions. Hence, honey having low protein content is better for storage and safe for consumption from different sugars (honey); most of these are proteins (Rahman *et al.*, 2017). About 20 different non-enzymatic proteins were reported; usually, different kinds of honey were 'albumins, globulins, proteases and nucleoproteins' (Doner, 2003).

Amino Acid

These are the organic complex found in honey and accountable for some important uses having antioxidant properties. The majority of amino acids present in the honey occurs only as part of a large construction and predigested form of AA (amino acid) can be less(one-fifth of the total). Among the different amino acids present in honey most common include 'proline, glutamic acid, alanine, phenylalanine, tyrosine, leucine and isoleucine' (Girolamo *et al.*, 2012). Besides this, other AAs recognised in lower concentrations in various varieties of honey are glutamine, histidine, 'glycine, threonine, arginine, valine, methionine, cysteine, tryptophan, lysine and serine' (Hermosın *et al.*, 2003; Kečkeš *et al.*, 2013). Among all AA (amino acid), CHNO (proline) is predominant in honey (Iglesias *et al.*, 2006). It is about 49% in liquid syrup (nectar honey) and 59% in fresh sugar (honeydew honey) among entire amino acids present. Proline arises from the saliva of honeybees while converting liquid into sugar (honey). When honey is stored for a long time, then amino acid content will be reduced, and then $C_5H_9NO_2$ (protein) helps to indicates honey ripeness (da Silva *et al.*, 2016). Besides, this is also advocated while measuring sugar (honey) adulteration with glucose due to honey concentration decreasing in sugar contaminated honey. The smallest value of proline in authentic honey is 180 mg/kg for CHNO that is accepted worldwide for the various honey verities (Hermosın *et al.*, 2003), except locust honey familiar to have minimum CHNO substances (Sakač *et al.*, 2019).

Enzymes

Enzymes are an important part of honey. Usually, natural honey is known to have enzymes in small amounts; the most important are diastase, invertase and glucose oxidase. Apart from these, other enzymes such as acid phosphatase, catalase, β-glucosidase, maltase, β-fructofuranosidase and ascorbic oxidases were also found in honey (Machado De-Melo *et al.*, 2017). Some of these enzymes were added by honeybees during the conversion of nectar to honey, i.e., glucosidase and enzyme (glucose oxidase) and therefore of bee origin (Doner, 2003). However, some enzymes like catalase and acid phosphatase have botanical origins from nectar, honeydew or pollen. Catalysts, e.g., 'diastase', form more the one source, i.e., from the hypopharyngeal gland as well as botanical origin. Another possible origin could be the plant sap feeder, which forms pepo (honeydew) (Machado De-Melo *et al.*, 2017). Storage conditions and decrystallisation control the concentration of protein in honey.

Diastase

Diastase (amylase) destroys starch while added by the bees during the ripening process. It is most stable among other enzymes and thermal resistors; hence, it is the most popular important parameter of honey novelty and diastase activity being regulated within the legislation, i.e., minimum value 8. It's determined that the number of alpha-amylase can change 0.01 g dextrose-directed deadline in 1 h (at 400°C), known as one unit of diastase activity. Consequence indicated in 'Schade units per gram of honey', known as DN (diastase number).

Invertase plays a significant role as a catalyst that transforms 'nectar and honeydew' sucrose into fructose and glucose. In addition to this, some oligosaccharides are also produced by the invertase and transglycosylase activity (Machado De-Melo *et al.*, 2017). Invertase activity continued even after the extraction of honey and its storage. Invertase activity of Indian honey is greater than European honey.

Glucose Oxidase

It oxidises a small amount of glucose to glucone, lactone and gluconic acids, increasing the acidity of honey and simultaneously releasing a small amount of hydrogen peroxide. This compound is responsible for the stability of ripening honey and provides microbial resistance. Normally, glucose oxidase is negligible in full density honey. In undiluted honey, the $C_6H_{12}O_7$ substance reduces hydrogen (pH) potential, therefore suppressing the enzyme action.

Catalase and Acid Phosphatase

Catalase decomposes the hydrogen peroxide produced through 'glucose oxidase' to H_2O and O_2, whereas acidic phosphatase produces PO_4^{3-} from PO_3^{-}. Both enzymes originate chiefly from nectar, pollen and honeydew. The value of phosphatase activity of honey can be related to honey fermentation. The pH of honey strongly influences it; the higher the PH, the higher the phosphatase activity. The optimum range for the activities of phosphatase lies between 4.5 and 6.5. Phosphatase activity is found to be higher in kinds of honey from oceanic climate (Alonso-Torre *et al.*, 2006). It is found in honey in a small range from 0.1 to 0.2% of the composition of nectar honey but can exceed 1% in another type of honey, such as honeydew honey. K, Na, Cl and Mg are major minerals present in honey, whereas Fe, Cu, Mn, Cl are a little liberal and B, P, S, Si and Ni are present in a small amount (Doner, 2003).

Vitamins

Honey contains fewer vitamins and hardly has any nutritional significance. These primarily come along with dust mite, nectar from flora (Sainz-Laın and Gomez-Ferreras, 2000). There is the presence of different vitamins such as 'ascorbic acid, pantothenic acid, niacin and riboflavin', thymine pyridoxine. The important vitamin is vitamin C having an antioxidant effect.

Phenolic Compound

The synthetic resin of honey carotenoid plays a significant role in 'aromatic secondary metabolites' of plant origin (Ferreres *et al.*, 1992). The level of these compounds in honey is 5–1300 mg/kg (Al-Mamary *et al.*, 2002; Gheldof and Engeseth, 2002). In honey, they are mostly phytochemical, phenylic acids and phenolic acid derivatives. Phenylic acid proportion in honey depends on the nectar content; it can be helpful to classify especially in the same floral diversity (Cianciosi *et al.*, 2018).

Volatile Components

Volatile component of honey contributes to the colour, flavour and taste. The important volatile components responsible for colour are flavonoids, tetraterpenoids, phylloxanthins and anthocyanin found together in water and lipid solvent colouration. Aroma and taste are important characteristics of honey and related to volatile compounds. More than 600 different substances can be reported in kinds of honey at extremely low consolidation. These are presented in honey as a complicated blend of distinct 'monoterpenes, terpenes, terpenoids, norisoprenoids, phenolic compounds, benzene derivatives, alcohols, ketones, aldehydes, esters, fatty acids, acids, hydrocarbons and cyclic compounds' that influence the aroma depending upon their concentration. The taste of honey is due to sugars, gluconic acid and proline in addition to volatile constituents (Machado De-Melo *et al.*, 2017).

Hydroxymethylfurfural (HMF)

Hydroxymethylfurfural (HMF) is an organic compound formed as a breakdown product of simple sugars (such as fructose), which is formed in a slow and natural process during the storage of honey, and much more quickly when honey is heated (Machado De-Melo *et al.*, 2017). HMF can also be formed from carbohydrates and glucose, giving $C_6H_{12}O_6$ overoxidisation from the oligo- and polysaccharides that can yield hexoses upon hydrolysis. It can also form during the Maillard reaction. Several factors affect the initiation of HMF, such as 'temperature, heating time, storage conditions, pH and honey type'. A high content of HMF, i.e., more than 100 mg/kg, can be an indicator of honey adulteration. So, CAC (Codex Aliment Arius Standard commission) defines an extreme level (limit) for HMF in honey, i.e., 40 mg/kg, to make sure about the heat level throughout preparation and also make sure about the safety of use. The amount of HMF production is related to fructose: glucose ratio, and it can be found at pH 4.6; monosaccharides (fructose) possess 5× times more susceptibility than $C_6H_{12}O_6$ and more the monosaccharides (fructose), the $C_6H_{12}O_6$ proportion has stimulated the activity (Shapla *et al.*, 2018).

Therapeutic Properties of Honey

Honey is one of the oldest remedies used to treat infections and microbial diseases since ancient time. Recent studies found that honey is the usual immunity booster fighting against many diseases (Erler and Moritz, 2016). Due to increased interest in the antimicrobial effectiveness of honey, antibiotics and new therapies are on the horizon. Recent findings show that honey has therapeutic property fight against microorganisms, i.e., virus, bacteria and fungi (Carter *et al.*, 2016). Even though bee produced honey worldwide, its properties depend upon their bee species, source of nectars (botanical source) and environmental factors. Other external factors include processing, storage condition and environmental conditions (Erler and Moritz, 2016; Oryan *et al.*, 2016). Due to huge changes in honey compounds, the therapeutic properties of honey are significantly multifaceted as a result of the mechanism of many compounds.

The therapeutic potential of honey is greatly complex as a result of the action of various compounds and large variations in the concentrations of these compounds among kinds of honey. The major biological properties used as a therapeutic agent are 'antimicrobial' (bactericidal or fungicidal), 'bacteriostatic' (or fungistatic), 'anti-inflammatory potential', 'radical scavenging activity' and 'antiviral activity' (Al-Waili *et al.*, 2011; Alam *et al.*, 2014; Nweze *et al.*, 2016). It can be used as an immunity booster, wound healing and skin disinfection. It also is used to treat medical conditions like diarrhoea, dermatitis, asthma, tumour, ulcer and diabetes. Honey is natural and nontoxic, a substitute or supplement to therapeutic agents but in some cases limit its use. The experimental applicability of honey has been lacking an exact mechanism for defining the activity of honey and differences of honey (Carter *et al.*, 2016; Meo *et al.*, 2017). Honey has strong effects and nontoxic, and some studies revealed that honey has great antibacterial properties on bacteria (at least 60 species) and other antibiotics.

The advantages of honey have been inscribed since ancient times by documented in ancient scriptures. The honey composition included glucose, fructose and minerals., Mg, Ca, K, NaCl, Fe and P. Honey contain vitamins (B1, B2, B3, B5, B6, C); it depends on the quality of nectar and pollen. Honey pH is 3.2–4.5 depending on the environmental factors and honey bee species. Honey is mainly composed of two monosaccharides. It prevents the growth of bacteria due to the acidic pH of honey. The mixture of water and honey loses its low water activity, and most water molecules are connected with sugars.

Antimicrobial Activity

Therapeutically, honey's importance of antimicrobial activity cannot be exaggerated, where the immune responses (body) may be insufficient to clear the infection. For those microorganisms which have developed resistance against antibiotics, honey has proven to be an effective antimicrobial property that provides resistance against both non-pathogenic and pathogenic microorganisms, i.e., yeasts, bacteria and fungi. The antimicrobial effect of honey might be inhibiting the growth of bacteria or capable of killing bacteria (Oryan *et al.*, 2016). Nevertheless, its potentials have been attributed to particular variables such as increase osmolarity (decrease water action), low acidic pH, H_2O_2 (hydrogen peroxide) and non-peroxide components (Almasaudi *et al.*, 2013; Alvarez-Suarez *et al.*, 2014). Fahim *et al.* (2014) studied that the acidity of honey prevents the growth of microbes, and its pH ranges between 4.0 and 4.5. Due to H_2O_2 activity and glucose oxidase activity, antimicrobial activity has been reported, especially in honey diluted form. The breakdown of H_2O_2 produces free radicals, which kills microorganisms. Therefore, due to catalase activity of honey, it could be terminated. Due to non-peroxide activity, this property of honey is more steady and stable, but sometimes the antibacterial activity of honey is a result of the peroxide effect. They are also called 'non-peroxide' kinds of honey, for example, such in Australia '*Leptospermum polygalifolium*' and '*Leptospermum scoparium*', which are assumed to have unknown dynamic constituents apart from the production of H_2O_2 (hydrogen peroxide). Other types of kinds of honey hold their effectiveness to inhibit microbes in the presence of catalase (Carter *et al.*, 2016; Oryan *et al.*, 2016).

It has been suggested the movement of honey and its origin partly due to plant origin. Organic solvents such as *n*-hexane, ethyl acetate, diethyl ether and chloroform are done using liquid-phase and solid-phase extraction methods. It could be used for the removal of compounds showing their activity. The detached mixtures have been accounted for to integrate flavonoids and other random mixes such as phenolic acids, proteins, lipids, ascorbic acids, carotenoid and natural acids (Kaur *et al.*, 2013; Julianti *et al.*, 2017). Other important effects of honey have prebiotic properties such as fructo-oligosaccharides or other types of oligosaccharides. Other oligosaccharides revealed an increase in bacteria, such as lactobacilli (intestinal microflora in human beings) (Cornara *et al.*, 2017). Some bacteria, such as lactobacillus, defence against some infection, i.e., salmonellosis, and other types of bacteria such as Bifidobacterium species, limit yeasts' overgrowth bacteria the gut wall (Finegold *et al.*, 2014; Cornara *et al.*, 2017).

In historical time, the use of honey was a conventional medicine for bacterial infections. Studies reported that the manuka honey had been the best-proven remedy against human pathogens such as *Enterobacter aerogenes*, *Escherichia coli*, *Staphylococcus aureus* and *Salmonella typhimurium*. Honey is effective against 'methicillin-resistant *S. aureus*', 'vancomycin-resistant enterococci' and streptococci (Carter *et al.*, 2016). Antimicrobial honey may have benefits over or resemblances with manuka honey (Alvarez-Suarez *et al.*, 2014). According to Fahim *et al.* (2014), *Pseudomonas aeruginosa* and *Enterococcus* species were more vulnerable, just like *S. aureus* (Fahim *et al.*, 2014; Almasaudi *et al.*, 2017).

Honey as a Natural Immune Booster

Honey has antibacterial properties, and to defeat the interlopers, honey is recognised as having no infection by 'immune system stimulation'. Recently, a substantial detail confirms that it is an innate immunity enhancer. Some recent reports have described the exciting activity of 'B60 lymphocytes and T lymphocytes' to raise the amount (number) in cell sample or also trigger neutrophil. Furthermore, in somatic culture, leucocyte may be excited to free 'cytokines IL-1, IL-6 or TNF-α'. These are cell

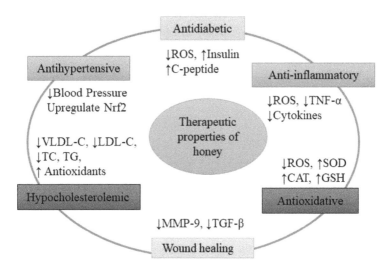

FIGURE 6.1 Showing therapeutic properties of honey.

carriers (transporter) that may trigger some immunologic responses to contamination (Israili, 2014). Carter *et al.* (2016), they negotiate the 'TNF-α production' in phagocyte to excite the 5.8 kDa. 'Manuka and pasture honey' were detected to enhance the lymphocytes originated from 'MM6 cells' where the value of P is greater than 0.001. Based on macrophage record, honey is a good derivative of $C_6H_{12}O_6$ which is necessary to the 'respiration burst' to lead to H_2O_2 production, the leading characteristic of their antibacterial action. And, honey is a major component of glucose production, or primary method in phagocyte to produce strength, or later properly function in damage cell and mucus wherever the O_2 transport becomes poor frequently. The acidic pH of honey plays an important role in the digestion of dead bacteria in the phagocytic method. Recently, it is concluded that in 'AIDS patients', the 'prostaglandin' value is increased when they use honey 80 g/day, which refines one 'immune system' (Figure 6.1) (Al-Waili *et al.*, 2011).

Anti-Inflammatory and Immunomodulatory Activities

On the other hand, it is proven that inflammatory infection is so dangerous that it can cause more damage. In the present day, the analysis of the activities is regulated in initiatory laboratory work 'animal models' and in cell refinement. The result of excessive inflammatory action can lead to 'free radical' formation in the cell. Such detached radicals are regulated through WBC that acts as significant exposure of inflammation reaction. It helps regulate the 'growth factor' that affects 'fibroblasts, angiogenesis or epithelial cells'. There are various types of honey in several other parts of the world, which shows anti-inflammation activities (Carter *et al.*, 2016; Oryan *et al.*, 2016). Phenolic compounds are the major content of honey by which it shows 'anti-inflammatory' property. Phenolic or flavonoid content is the source of inhibition of pro-inflammatory activities of iNOS (inducible nitric oxide synthase), cyclooxygenase-1(COX-1) and cyclooxygenase-2 (COX-2). The 'prostaglandins concentration' reduced when organic honey can be absorbed (Al-Waili *et al.*, 2011). Therefore, honey acts as an anti-inflammatory, and it can cure the 'corticosteroids and NSAIDS' problems.

It can also regulate the protein action like 'inclusive of iNOS, cyclooxygenase-2, tyrosine kinase and ornithine decarboxylase'. The initiation of formation of 'tumor necrosis alpha' has been reported earlier in various forms of honey, i.e., IL-6, and IL-1β. 'Gelam honey' is responsible for the decrease in the inflammatory action, e.g. 'TNF-α and COX-2' which may lead to inhibition the induction of 'NF-Kb pathway'. Also, NF-Kb stimulation can play an important role to cause inflammatory diseases. The formation of SCFA (short-chain fatty acid) may lead to obstructing ingestion of honey, and SCFA also is responsible for the 'immunomodulatory activities'. Honey sugars can affect 'immune response'. Also, there are no sugar constituents that are mainly liable to immunomodulation activities (Alvarez-Suarez *et al.*, 2014). The conclusion is that honey also has 'anti-immunomodulatory and anti-inflammatory activities'.

Role of Honey in Cosmetic Industry

Honey is used in cosmetics since ancient time. Also, in the present day, honey is used in various cosmetics. It is a constituent of dissolvable cosmetic amalgam as a humectant towards beauty care products or skin. Overall, these skincare products are appropriate for any skin. It (Honey) has properties like absorptive, antiseptic or antifungal, and constituents can nourish the skin. Its slightly acidic nature boosts the upper layer, where the pH of the skin is about 5.5. Skincare products such as 'shampoo', 'hair balm', 'purifying lotion with honey', 'hand cream' and 'sun cream' are great products with honey texture. Honey sustains our skin, or it is a good hydrator. If we use it daily, skin becomes young or reduces wrinkle emergence (Bogdanov, 2016). The production of these products (cosmetic ingredients) was determined based on the US FDA (Food and Drug Administration) facts or cosmetic industry through which these constituents are utilised. The product's prevalence found in the products was accumulated from the producers or through the skincare products section in Food and Drug Administration VCRP (Voluntary Cosmetic Registration Program) information.

A survey in 2019 Voluntary Cosmetic Registration Program (VCRP) details, which conclude that about 1002 preparations occur by this (honey). And about 359 preparations form through the tincture of Honey. About 22% of honey is used to make paste mask or 'mud packs'. About 7% of honey tincture has appeared in body and hand products. The information is described for 'honey Cocoats' by the council, and on the other hand, there are no such data discussed by VCRP. VCRP has data only for 'honey powder' but its consolidation is not described. 'Hydrolysed honey' and 'hydrolysed honey proteins' are not utilised as per Voluntary Cosmetic Registration Program or management. Honey is also used in 'baby products' (which are used near eyes). About 0.01% honey is determined in 13 'baby products', 3% in 20 'lipsticks', 0.00035% in dentifrices, 0.1% in oral hygiene and others.

The tincture of honey is utilised in about 14 bath products, or which forms 'mucous membrane exposure'. However, honey and its tincture are also used in 'cosmetic sprays'. Also, up to 3% honey is utilised to prepare 'face powder' (Aylott *et al.*, 1979; Russell *et al.*, 1979). And these constituents are considered in the 'European Union inventory of cosmetic ingredients'. Honey is good for healthy skin. There are many reports found through which it considered that so many cosmetic products have honey or other bees' products (Krell, 1996). The variability of honey utilised is according to the type of product. Normally, honey is used in small quantity in the manufacturing of goods, 'creams' and high quantity in 'anhydrous ointments' about 10–15%. A conditioning agent is determined good hair, strengthening the 'hair elasticity' or 'flexibility'. For the production of cosmetics and soaps, fermented, dehydrated honey is utilised. The average utilisation of honey in cosmetics is 1–10%.

Honey as Moisturiser

In various research works, honey may be considered a protector or may relieve skin due to its composition. For the nourishment of the skin, a moisturiser is formulated (Burlando and Cornara, 2013). It is used to increase skin functioning by treating skin dryness (Eady *et al.*, 2013). Mostly the moisturisers are used according to the kind of skin and its conditions. Mostly 'emollients', 'humectants' and some other constituents are used to prepare moisturisers. Honey's humectant properties are responsible for acting as a moisturiser (Eady *et al.*, 2013; Hadi *et al.*, 2016). However, the required process is undetermined. It is presumed that humectant honey has a good quantity of glucose and fructose ratio. They create H+ bridges with H_2O, which keeps humidity in aroused skin covering and allocates moisturising to the skin. This effect does not come from fructose and glucose levels, also obtained from many other constituents such as amino acid $C_5H_9NO_2$, in addition to it, 'arginine, alanine, glutamic acid, aspartic acid, lysine, glycine and leucine (amino acid) or largely gluconic acid and, to a certain extent, lactic, citric, succinic, formic, malic, acetic, malic and oxalic acids' (organic acid) which increase the effect of 'glucose' or 'fructose' to keep the skin hydrated (Burlando and Cornara, 2013).

Healthy Hair

A soft, smooth, glossy or flexible are the features of healthy hair that can resist shave and strip strength, i.e., abrasion (Greenwood and Handsaker, 2012). The systematic and somatic features of hair are based upon hair shape having 'cuticles' and 'cortex' (Sinclair, 2007). Hair stringiness and shine or lustre based on cortex and cuticle part, respectively (Robbins and Crawford, 1991). Vanguard for self-protection from water is the first layer, i.e., epicuticle. The hair colour depends upon the appearance of 'eumelanin' and 'pheomelanin' for red and black/brown. Damage will occur because of insufficient care and some climate integrant and too by synthetic processes (Marsh *et al.*, 2015). For better hair, we should clean hair properly. It is hence proved that 'shampoo' and 'conditioners' having hone has useful results to maintain hairs (Paus and Cotsarelis, 1999). The major usefulness of honey is sticking into hair completely, which makes hair shiny and soft (Burlando and Cornara, 2013). There are various useful properties of honey having medicinal use also because of its 'antimicrobial effect'. H_2O_2, pressure level, huge acidic level, $C_6H_{12}O_6$ lysozyme and flavonoids are the factor which forms it (Bogdanov, 1997). It helps to reduce bacterial and fungal development in the body. Honey is useful to prevent wound healing and in various therapies like Hebra disease, dermatophytosis, seborrheic dermatitis, furfur, nappy infection,

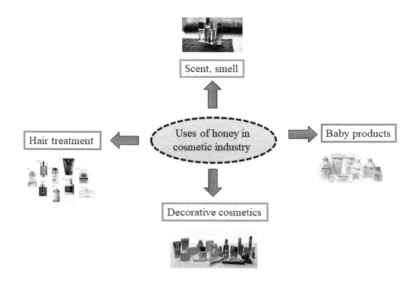

FIGURE 6.2 Showing uses of honey in a cosmetic industry.

skin disease, cystitis and rectal slit (Burlando and Cornara, 2013). In manuka honey, there are antibacterial properties in a higher amount than that of other forms of honey, which consist of 'methylglyoxal' than European honey. It also cures gingivitis and prevents the growth of oral biofilm, 'fight thrush' and meningitis. Substituted phenyl is also a reason for the bacterial properties of various honey collection. Due to its huge nourishing properties, honey is also utilised in beauty treatment. It has carbohydrates, fruit acid or various trace elements present in large quantities that are liable for nutritive or restoration efficacy.

Moreover, it is a good absorbent and good cleaner of the dermal tissue. It can cure wrinkled skin, shininess, smoothening and colour of skin. Fruit acids help to form moulting and moisturise the dead skin. A sugared honey acts as a catalytic agent. Vitamins are present in honey which can also penetrate easily in the skin. Dryness (abnormal dryness) of skin is soothed through 'fatty acid' and 'mineral salt' present in honey. Honey has sugar content, some oils and 'bio-elements' by which its properties increase and used in various cosmetics like cracked creams, balms, bathing products etc. due to its tint, soften, strengthening consequence. Also, honey consists of flavonoids constituents which are useful to cure skin annoyance (Figure 6.2) (Kurek-Górecka *et al.*, 2020).

REFERENCES

Abdulkhaliq, A. and Swaileh, K.M. 2017. Physico-chemical properties of multi-floral honey from the West Bank, Palestine. *International Journal of Food Properties*, 20(2): 447–445.

Ahmed, S., Sulaiman, A.S., Baig, A. A., Ibrahim, M., Liaqat, S., Fatima, S., Jabeen, S., Shamim, N. and Othman, H. N. 2018. Quality characteristics of honey: a review. *Oxidative Medicine and Cellular Longevity*, 19: 8367846.

Ajibola, A. 2015. Physico-chemical and physiological values of honey and its importance as a functional food. *International Journal of Food Sciences and Nutrition*, 2(2): 180–188.

Alam, F., Islam, M.D., Gan, S.H. and Khalil, M., 2014. Honey: a potential therapeutic agent for managing diabetic wounds. *Evidence-Based Complementary and Alternative Medicine*.

Al-Mamary, M., Al-Meeri, A. and Al-Habori, M. 2002. Antioxidant activities and total phenolics of different types of honey. *Nutrition Research*, 22: 1041–1047.

Almasaudi, S.B., Al-Nahari, A.A., El Sayed, M., Barbour, E., Al Muhayawi, S.M., Al-Jaouni, S., Azhar, E., Qari, M., Qari, Y.A. and Harakeh, S., 2017. Antimicrobial effect of different types of honey on *Staphylococcus aureus*. *Saudi Journal of Biological Sciences*, 24: 1255–1261.

Almeida-Muradian, L.B. de, Stramm, K.M., Leticia, M. and Estevinho, L.M. 2014. Efficiency of the FT-IR ATR spectrometry for the prediction of the physicochemical characteristics of *Melipona subnitida* honey and study of the temperature's effect on those properties. *International Journal of Food Science and Technology*, 49: 188–195.

Almeida-Muradian, L.B., Stramm, K.M., Horita, A., Barth, O.M., Freitas, A.S. and Estevinho, L.M. 2013. Comparative study of the physicochemical and palynological characteristics of honey from Melipona subnitida and *Apis mellifera*. *International Journal of Food Science and Technology*, 48: 1698–1706.

Alonso-Torre, S.R., Cavia, M.M., Fernández-Muiño, M.A., Moreno, G., Huidobro, J.F. and Sancho, M.T. 2006. Evolution of acid phosphatase activity of honeys from different climates. *Food Chemistry*, 97: 750–755.

Alvarez-Suarez, J.M., Gasparrini, M., Forbes-Hernández, T.Y., Mazzoni, L. and Giampieri, F. 2014. The composition and biological activity of honey: a focus on manuka honey. *Foods*, 3: 420–432.

Al-Waili, N., Salom, K. and Al-Ghamdi, A.A. 2011. Honey for wound healing, ulcers, and burns; data supporting its use in clinical practice. *The Scientific World Journal*, 11: 766–787.

Aylott, R.I., Byrne, G.A., Middleton, J.D. and Roberts, M.E. 1979. Normal use levels of respirable cosmetic talc: preliminary study. *International Journal of Cosmetic Science*, 3: 177–186.

Ball, W. D. 2007. The chemical composition of honey. *Journal of Chemical Education*, 84(10): 1643–1646.

Bartulović, M. 2015. *Proizvodnja i plasman meda OPG-a Antunović*. Diplomski rad, Poljoprivredni fakultet – Osijek, Sveučilište Josipa Jurja Strossmayera, Osijek.

Bertoncelj, J., Golob, T., Kropf, U. and Korosec, M. 2011. Characterisation of Slovenian honeys on the basis of sensory and physicochemical analysis with a chemometric approach. *International Journal of Food Science and Technology*, 46: 1661–1671.

Bogdanov, S. 2011a. Honey technology. In S. Bogdanov (Ed.), *The honey book*, 15–18.

Bogdanov, S. 2011b. Physical properties. In S. Bogdanov (Ed.), *The honey book*, 19–27.

Bogdanov, S. 1997. Nature and origin of the antibacterial substances in honey. *LWT-Food Science and Technology*, 30: 748–753.

Bogdanov, S. 2016. Beeswax: History, Uses and Trade. *Bee Product Science*. http://www.bee-hexagon.net/

Bogdanov, S., Lullman, C., Martin, P., Von der Ohe, W., Russmann, H., Vorwohl, G. and Vit, P. 1999. Honey quality and international regulatory standards: Review by the International Honey Commission. *Bee World*, 80: 61–69.

Burlando, B. and Cornara, L. 2013. Honey in dermatology and skin care: A review. *Journal of cosmetic dermatology*, 12: 306–313.

Carter, D.A., Blair, S.E., Cokcetin, N.N., Bouzo, D., Brooks, P., Schothauer, R. and Harry, E.J. 2016. Therapeutic manuka honey: no longer so alternative. *Frontiers in Microbiology* 7: 569.

Cavia, M.M., Fernandez-Muino, M.A., Huidobro, J.F. and Sancho, M.T. 2004. Correlation between moisture and water activity of honeys harvested in different years. *Journal of Food Science*, 69: 368–370.

Cianciosi, D., Forbes-Hernández, T.Y., Afrin, S., Gasparrini, M., Reboredo-Rodriguez, P., Manna, P.P., Zhang, J., Lamas, L.B., Flórez, S.M., Toyos, P.A., Quiles, J.L., Giampieri, F. and Battino, M. 2018. Phenolic compounds in honey and their associated health benefits: A review. *Molecules* 2018 Sep; 23(9): 2322.

Conforti, P.A., Lupano, C.E., Malacalza, N.H., Verónica, A. and Castells, C.B. 2006. Crystallization of Honey at −20°C. *International Journal of Food Properties*, 9(1): 99–107.

Cornara, L., Biagi, M., Xiao, J. and Burlando, B. 2017. Therapeutic properties of bioactive compounds from different honeybee products. *Frontiers in Pharmacology*, 8: 412.

Costa, P.A., Moraes, I.C.F., Bittante, A.M.Q.B., Sobral, P.J.A., Gomide, C.A. and Carrer, C.C. 2013. Physical properties of honeys produced in the Northeast of Brazil. *International Journal of Food Studies*, 2, 118–125.

Crane, E.E. 1980. *A book of honey*. Oxford: Oxford University Press. ISBN 9780192860101.

Czipa, N., Borbély, M. and Győri, Z. 2012. Proline content of different honey types. *Acta Alimentaria*, 41(1): 26–32.

da Silva, P.M., Gauche, C., Gonzaga, L.V., Costa, A.C.V. and Fett, R. 2016. Honey: Chemical composition, stability and authenticity. *Food Chemistry*, 196 (2016): 309–323.

Del Nozal, M.J., Bernal, J.L., Marinero, P., Diego, J.C., Frechilla, J.I., Higes, M. and Llorente, J. 1998. High performance liquid chromatographic determination of organic acids in honeys from different botanical origin. *Journal of Liquid Chromatography & Related Technologies*, 21: 3197–3214.

Dinkov, D. 2003. A scientific note on the specific optical rotation of three honey types from Bulgaria. *Apidologie*, 34: 319–320.

Do Nascimento, S.A., Marchini, C.L., de Carvalho, L.A.C., Araújo, D.F.D., de Olinda, A.R. and da Silveira A.T. 2015. Physical-chemical parameters of honey of stingless bee (Hymenoptera: Apidae). *American Chemical Science Journal*, 7(3): 139–149.

Doner, L.W. 2003. Honey. In B. Caballero, P.M. Finglas, & L.C. Trugo (Eds.), *Encyclopedia of food sciences and nutrition*, 3125–3130. London: Academic Press.

Eady, E.A., Layton, A.M. and Cove, J.H. 2013. A honey trap for the treatment of acne: manipulating the follicular microenvironment to control *Propionibacterium acnes*. *BioMed Research International*, 2013;2013:679680. DOI: 10.1155/2013/679680. Epub 2013 May 14. PMID: 23762853; PMCID: PMC3666392.

Erler, S. and Moritz, R.F.A. 2016. Pharmacophagy and pharmacophory: mechanisms of self-medication and disease prevention in the honeybee colony (*Apis mellifera*). *Apidologie* 47: 389–411.

Fahim, H., Dasti, J.I., Ali, I., Ahmed, S. and Nadeem, M. 2014. Physico-chemical analysis and antimicrobial potential of *Apis dorsata*, *Apis mellifera* and *Ziziphus jujube* honey samples from Pakistan. *Asian Pacific Journal of Tropical Biomedicine*, 4: 633–641.

Ferreres, F., Ortiz, A., Silva, C., García-Viguera, C., Tomás-Barberán, F.A. and Tomás-Lorente, F. 1992. Flavonoids of "La Alcarria" honey: A study of their botanical origin. *Zeitschrift für Lebensmittel-Untersuchung und -Forschung*, 194: 139–143.

Finegold, S.M., Li, Z., Summanen, P.H., Downes, J., Thames, G., Corbett, K., Dowd, S., Krak, M. and Heber, D. 2014. Xylooligosaccharide increases bifidobacteria but not lactobacilli in human gut microbiota. *Food & Function*, 5: 436–445.

Garcıa-Álvarez, M., Ceresuela, S., Huidobro, J.F., Hermida, M. and Rodríguez-Otero, J.L. 2002. Determination of polarimetric parameters of honey by near-infrared transflectance spectroscopy. *Journal of Agricultural and Food Chemistry*, 50: 419–425.

Gheldof, N. and Engeseth, N.J. 2002. Antioxidant capacity of honeys from various floral sources based on the determination of oxygen radical absorbance capacity and inhibition of in vitro lipoprotein oxidation in human serum samples. *Journal of Agricultural and Food Chemistry*, 50: 3050–3055.

Girolamo, F.D., D'amato, A. and Righetti, P.G. 2012. Assessment of the floral origin of honey via proteomic tools. *Journal of Proteomics*, 75: 3688–3693.

Gomes, S., Dias, L., Moreira, L., Rodrigues, P. and Estevinho, M.L. 2009. Physicochemical, microbiological and antimicrobial properties of commercial honeys from Portugal. *Food and Chemical Toxicology*, 48: 544–548.

Greenwood, M. and Handsaker, J. 2012. Honey and Medihoney Barrier Cream: Their role in protecting and repairing skin. *British Journal of Community Nursing*, 17: 32–37.

Hadi, H., Syed Omar, S.S. and Awadh, A.I. 2016. Honey, a gift from nature to health and beauty: A review. *British Journal of Pharmacy*, 1: 46.

Haroun, M.I., Poyrazoglu, E.S., Konar, N. and Artik, N. 2012. Phenolic acids and flavonoids profiles of some Turkish honeydew and floral honeys. *Journal of Food Technology*, 10: 39–45.

Hermosın, I., Chicoón, R.M. and Cabezudo, M.D. 2003. Free amino acid composition and botanical origin of honey. *Food Chemistry*, 83: 263–268.

Iglesias, M.T., Martian-Alvarez, P.J., Polo, M.C., Lorenzo, C., Gonzalez, M. and Pueyo, E.N. 2006. Changes in the free amino acid contents of honeys during storage at ambient temperature. *Journal of Agricultural and Food Chemistry*, 54: 9099–9104.

Israili, Z.H. 2014. Antimicrobial properties of honey. *American Journal of Therapeutics*, 21: 304–323.

Jakubik, A.P., Borawska, M.H. and Socha, K. 2020. Modern methods for assessing the quality of bee honey and botanical origin identification. *Food*, 9: 1028.

Julianti, E., Rajah, K.K. and Fidrianny, I. 2017. Antibacterial activity of ethanolic extract of cinnamon bark, honey, and their combination effects against acne-causing bacteria. *Scientia Pharmaceutica*, 85: 19.

Kaur, R., Kumar, N.R. and Harjai, K. 2013. Phytochemical analysis of different extracts of bee pollen. *International Journal of Pharmaceutical and Biological Research*, 4: 65–68.

Kečkeš, J., Trifković, J., Andrić, F., Jovetić, M., Tešić, Z. and Milojković-Opsenica, D. 2013. Amino acids profile of Serbian unifloral honeys. *Journal of the Science of Food and Agriculture*, 93: 3368–3376.

Krell, R. 1996. Value-added products from beekeeping. *Food and Agriculture Services Bulletin*, 124.

Krell, R. 1996. *Value-added products from beekeeping* (No. 124). Food & Agriculture Org.

Kurek-Górecka, A., Górecki, M., Rzepecka-Stojko, A., Balwierz, R. and Stojko, J. 2020. Bee products in dermatology and skin care. *Molecules*, 25: 556.

Machado De-Melo, A.A., Almeida-Muradian, L.B. de, Sancho, M.T. and Maté, A.P. 2017. Composition and properties of *Apis mellifera* honey: a review. *Journal of Apicultural Research*, 1–33.

Manyi-Loh, C.E., Clarke, A.M. and Ndip, R.N. 2011. An overview of honey: therapeutic properties and contribution in nutrition and human health. *African Journal of Microbiology Research*, 5: 844–852.

Marsh, J.M., Davis, M.G., Lucas, R.L., Reilman, R., Styczynski, P.B., Li, C., Mamak, M., McComb, D.W., Williams, R.E.A., Godfrey, S. and Chechik, V. 2015. Preserving fibre health: reducing oxidative stress throughout the life of the hair fibre. *International Journal of Cosmetic Science*, 37: 16–24.

Mato, I., Huidobro, J.F., Simal-Lozano, J. and Sancho, M.T. 2003. Significance of nonaromatic organic acids in honey. *Journal of Food Protection*, 66, 2371–2376.

Maurizio, A. 1962. From the raw material to the finished product: Honey. *Bee World*, 43, 66–81.

Meo, S.A., Al-Asiri, S.A., Mahesar, A.L. and Ansari, M.J. 2017. Role of honey in modern medicine. *Saudi Journal of Biological Sciences*, 24: 975–978.

Nayik, G.A., Dar, B.N. and Nanda, V. 2018. Rheological behavior of high altitude Indian honey varieties as affected by temperature. *Journal of the Saudi Society of Agricultural Science*, 17: 323–329.

Nikolova, K., Panchev, I., Sainov, S., Gentscheva, G.E. and Ivanova, E. 2012. Selected physical properties of lime bee honey in order to discriminate between pure honey and honey adulterated with glucose. *International Journal of Food Properties*, 15: 1358–1368.

Nweze, J.A., Okafor, J.I., Nweze, E.I. and Nweze J.E. 2016. Comparison of antimicrobial potential of honey samples from *Apis mellifera* and two stingless bees from Nsukka, Nigeria. *Journal of Pharmacognosy and Natural Products*, 2: 1–7.

Oroian, M. 2013. Measurement, prediction and correlation of density, viscosity, surface tension and ultrasonic velocity of different honey types at different temperatures. *Journal of Food Engineering*, 119: 167–172.

Oryan, A., Alemzadeh, E. and Moshiri, A. 2016. Biological properties and therapeutic activities of honey in wound healing: a narrative review and meta-analysis. *Journal of Tissue Viability*, 25: 98–118.

Paus, R. and Cotsarelis, G. 1999. The biology of hair follicles. *New England Journal of Medicine*, 341: 491–497.

Piazza, M.G. and Persano Oddo, L. 2004. Bibliographical review of the main European unifloral honeys. *Apidologie*, 35: S94–S111.

Rahman, M.M., Karmoker, P. and Alam, M.Z. 2017. Quality evaluation of some selected commercial honey products available in the market of Bangladesh. *Fundamental and Applied Agriculture*, 2(3): 326–330.

Robbins, C.R. and Crawford, R.J. 1991. Cuticle damage and the tensile properties of human hair. *Journal of the Society of Cosmetic Chemists*, 42: 59–67.

Russell, W.L., Kelly, E.M., Hunsicker, P.R., Bangham, J.W., Maddux, S.C. and Phipps, E.L. 1979. Specific-locus test shows ethylnitrosourea to be the most potent mutagen in the mouse. *Proceedings of the National Academy of Sciences*, 76: 5818–5819.

Sabatini, A.G. 2007. Il miele: Origine, composizione e proprietá. In A.G. Sabatini, L. Botolotti, & G.L. Marcazzan (Eds.), *Conscere il miele*, 3–37. Bologna-Milano: Avenue Media.

Sainz-Laın, C. and Gomez-Ferreras, C. 2000. Mieles Espanolas. Caracterısticas e identificación mediante el analisis de pollen Spanish honeys. In *Characteristics and identification by pollenanalysis*. Madrid: Mundi-Prensa.

Sakač, M.B., Jovanov, P.T., Marić, A.Z., Tomičić, Z.M., Pezo, L.L., Hadnađev, T.R.D., Novaković, A.R. 2019. Free amino acid profiles of honey samples from Vojvodina (Republic of Serbia). *Food and Feed Research*, 46 (2), 179–187.

Sanz, M.L., Gonzalez, M.M., & Martınez-Castro, I. 2002. Los azúcares de la miel [The sugars of honey]. In C. De Lorenzo (Ed.), *La miel de Madrid*, 95–108. Madrid: Consider a Economy's Innovation Technology' gica.

Shapla, U.M., Md. Solayman, M., Alam, N., Khalil, M.I., and Gan, S.H. 2018. 5-Hydroxymethylfurfural (HMF) levels in honey and other food products: effects on bees and human health. *Chemistry Central Journal*, 35: 1–18.

Silva, M.S., Rabadzhiev, Y., Eller, M.R., Iliev, I., Ivanova, I. and Santana, W.C. 2017. Microorganism in honey. In *Honey analysis*, 233–258. Intech Open, London, DOI: 10.5772/67262.

Sinclair, R.D. 2007. Healthy hair: what is it? *Journal of Investigative Dermatology Symposium Proceedings*, 12(2): 2–5.

Sohaimy, S.A. El, Masry, S.H.D. and Shehata, M.G. 2015. Physicochemical characteristics of honey from different origins. *Annals of Agricultural Science*, 60(2): 279–287.

Tong, Q., Zhang, X., Wu, F., Tong, J., Zhang, P. and Zhang, J. 2010. Effect of honey powder on dough rheology and bread quality. *Food Research International*, 43: 2284–2288.

Val, A., Huidobro, J.F., Sanchez, M.P., Muniategui, S., Fernández-Muino, M.A. and Sancho, M.T. 1998. Enzymatic determination of galactose and lactose in honey. *Journal of Agricultural and Food Chemistry*, 46, 1381–1385.

White, Jr., J.W. 1975. La miel [Honey]. In D. E. Hijos (Ed.), *La colmena y la abeja melıfera*, 397–428. Hamilton: Editorial Hemisferio Sur.

7

Honey Toxicity and Its Health Hazards Along with Related Mechanisms

Ashish Kumar Lamiyan, Ramkesh Dalal, Sapna Katnoria, Ahsan Ali and Neelima R. Kumar
Panjab University, Chandigarh, India

Anoop Singh
Postgraduate Institute of Medical Education and Research (PGIMER), Chandigarh, India

CONTENTS

Introduction

Honey may contain many poisonous substances which may have harmful effects on human health; it should not be believed to be a fully healthy product. Due to the heating process, poor packaging, poor handling and high humidity, some of these compounds may be generated. The source of these chemical compounds may be nectar obtained from toxic plants and from places that contain environmentally harmful substances (Islam *et al.*, 2014). Heavy metals, pesticides,

DOI: 10.1201/9781003175964-7

antibiotics and hazardous chemicals from industrial effluents render the honey toxic. Hydroxymethyl-furfural (HMF) has also received a large amount of attention among the toxic compounds in honey, and researchers need to concentrate on underlying mechanisms of modifications, from sugar to HMF so that its production stays minimal. Carcinogenic compounds such as furan-3-carboxylic acid (3-furoic acid), 2-furaldehyde (furfural), furan-3-carboxaldehyde (3-furaldehyde) and 2-aminobenzoic acid, furan-2-carboxylic acid (2-furoic acid), methyl ester have also been identified in addition to HMF (methyl anthranilate) in honey. To decide how and when toxicity in honey develops and how to decrease its production, a thorough investigation of toxic honey compounds is importantly needed. Also, gelsemine, tutin, pyrrolizidine alkaloids, hyoscine, GTXs, hyoscyamine, hyenanchin, oleandrin, saponin, strychnine and oleandrigenin can be probable toxins present in honey (Jansen *et al.*, 2012; Ullah *et al.*, 2018). In the world, honey adulteration has become a widespread occurrence, and more testing is needed to regulate the quality and marketing of honey. Honey also includes highly harmful heavy metals that may be present due to environmental pollution (such as As, Ni, Co, Se, Cr, Cd, Pb and Hg) (Solayman *et al.*, 2016). Therefore, far areas are the most suitable locations for honey production as the chances of pollution in environment are very less; moreover, it is necessary to maintain an effective quality control procedure for the marketing of honey (Grigoryan, 2016).

In order to safeguard human welfare, health authorities in all nations have to enact firm regulations and laws that control and regulate the processing, handling and storage of honey.

Types of Honey Toxicity

Metals

The nutritional properties of honey are affected by its maturation period, season, species of bees, storage vessel and source of nectar. Generally, honey contains only little amount of metals. The selection of flower and collection of nectar, pollen grain and honeydew by honeybees is largely affected by the amount and type of inorganic elements. Various types of honeys contain different types of metals and their concentration depends on the geographical location and nectar composition. Contamination by metals mainly comes when honeybees visit water sources polluted by industrial waste and industry emission.

It is also contributed in small amount by inappropriate method of handling and maintenance. Metal-contaminated honey may cause serious ill effects on human health if consumed regularly. Floral source of nectar may have biomagnified levels of heavy metals. During the last decade, honey contamination with metals was assessed in countries like China, Italy, France, Croatia, Slovenia, Poland and Turkey. These reports claimed the heavy metals not only in honey but also other bee products. Pollution from smelting industries not only spoiled drinking water and fertile land but also adversely affected human health and life of other animals if they consumed the natural product such as honey which contained accumulated amount of these elements (Kılıç Altun *et al.*, 2017).

The data from various honey samples revealed that zinc and copper were the metals which were found in higher concentration while lead, cadmium, nickel and chromium were present in small amounts. The presence of those metals in honey as highly influenced by botanical origin of element, geographical location and geochemical difference is found in honey sample taken from various locations (Bartha *et al.*, 2020).

From one trophic level to next trophic level, the metals accumulate and become highly toxic when they exceed maximum tolerance limit of human body. If honey contains metals above-mentioned permissible level, then it may cause adverse health issues, for example, honey contained high amount of copper and iron if consumed regularly with staple food causes gastrointestinal illness. If these heavy metals accumulated in human body at a toxic level, they may cause many serious health problems such as cancer, infertility, epilepsy and neurological disorder.

Some commonly found metals in honey include potassium, phosphorous, magnesium, aluminium, calcium, sodium, iron, manganese, copper, zinc, sulphur and selenium. Studies have shown that honey collected from sources near industrial areas contains heavy metals and may cause health hazards but there is also evidence that highly toxic metals such as lead, mercury and cadmium are found in very negligible amount as compared to other metals.

Honeybees may also collect heavy metal residues from flowers during foraging. The pollutant particles settle on flowers from the atmosphere and stick to the hair of honeybees with pollen grain collected honeybee the bees. Genetically engineered bees which do not contain hair on legs and body may reduce the accumulation of heavy metals in bee products but it would eliminate a very important function of honeybees that is pollination and that is a price man cannot afford to pay.

Other man-made sources by which honey can get contaminate dare volcanic smog, bush fire etc. Extensive use of pesticides and fertilisers may cause high levels of cadmium in soil which affects the overall ecosystem. The establishment of apiaries near the large commercial farming sites where agrochemicals, fertilisers containing heavy metals, fumigation of insects are practised causes contamination of honey and other bee products and these toxic chemicals also get accumulated in the bee hive.

A recent study demonstrated that unsuitable storage and improper processing could lead to contamination of honey. Mainly the impurities of zinc and other heavy metals in honey come from unmanaged storage, unhygienic handling, improper processing and use of galvanised vessel and tools.

In modern times, chronic health problems of nervous system, kidney, liver and respiratory systems are caused by heavy metals. Almost all types of heavy metals are highly carcinogenic at high concentration. Body growth and development is inhibited by regular intake of heavy metals in staple food, which is also responsible for imbalance in thermoregulatory system, metabolic disorder, and neurodegenerative diseases such as Parkinson's and Alzheimer diseases. In some conditions, heavy metals accumulate in human body and patient's immune system attacks on his/her own cells that lead to autoimmune disease. It causes many serious health problems such as rheumatoid arthritis, kidney failure, circulatory problem and hypertension (Matin *et al.*, 2016; Sall *et al.*, 2020).

The effects of some heavy metals on health along with their implications are discussed in the following sections.

Arsenic

When an individual is exposed to arsenic, it results in toxicity issues either acute or chronic. Acute poisoning results in blood vessels and stomach conditions being damaged. Chronic toxicity (arsenicosis) also refers to skin symptoms such as keratosis and pigmentation. When exposure is at lower levels it causes vomiting, nausea, reduced production of blood cells. Also, damage to blood vessels occurs which causes abnormal heartbeat and pricking sensation in body. Exposures for longer time cause lesions in skin, pulmonary, neurological and cardiovascular problems. In the case of chronic arsenicosis, changes occur in vital organs of the body which finally leads to death. The development of a number of cancers, including lung, liver, colon, skin, bladder, and kidney cancer, has also been reported due to arsenic exposure (Sall *et al.*, 2020).

Lead

In children and adults, the burden of lead in the human bloodstream complicates the gastrointestinal tract and central nervous system. The after-effects which follow lead accumulation are termed as lead poisoning. In case of exposure to lead, symptoms like sleeplessness, headache, renal dysfunction, abdominal pain, fatigue, hallucinations, vertigo, kidney damage, brain damage, mental retardation, birth defects, dyslexia, allergies, weight loss and even coma have been reported in countable cases. As it disturbs the working of most of the body organs, the danger of its impact remains a matter of concern. When exposure exceeds permissible limits, lead's elevated levels bring blood-brain barrier to interstitial spaces resulting in oedema. Similarly, it is seen that lead has impact on CNS as it possesses the potential to disrupt the intracellular second messenger system of the human body. Many prospective epidemiologic studies which took into consideration children less than 5 years of age reported impairment of intellectual development at low levels of lead exposure (Altunatmaz *et al.*, 2018).

Mercury

When mercury crosses the levels of bearable limits, it creates a lot of problematic issues to the biological system of humans. Sometimes, it is found to combine with other elements resulting in the formation of organic and inorganic forms of mercury. The different levels of exposure impose negative effects on kidney, brain and developing fetus. In some cases, it leads to development of carcinoma. The penetrating potential of organic mercury due to its lipophilic nature capacitates it to effect nervous system and cause alterations in brain functions. The major challenge in exposure to mercury is related to diagnosis of mercury poisoning because all the resulting symptoms are commonly shared with other illnesses such as vomiting, nausea, depression, fatigue skin rashes, diarrhoea, tremors, headache and high blood pressure. Also, in recent studies, it has been confirmed that during the period of pregnancy if women gets exposed to mercury through dietary intake, development in loss of memory, impaired speech, reduced motor neuron function and neural transmission in the offspring is seen.

Cadmium

In humans, there are multiple harmful health impacts due to cadmium and its derivatives. The health risks of cadmium toxicity are exacerbated because human body does not possess enough potential to secrete cadmium outside the body. The kidney, in particular, reabsorbs cadmium, thereby limiting its excretion. Also, cadmium inhalation for short term can cause significant lung damage and respiratory irritation when it is absorbed at higher dose. Long-term exposure in the bones and lungs to cadmium makes its accumulation higher and can cause bone and lung injury. It has been observed that higher levels of toxicity due to cadmium accumulation result in decreased bone density and as a result increase bone fracture risks for both male and female individuals. Due to higher deposition in proximal tubules in higher concentrations, cadmium possesses toxicity to renal system. Thus, cadmium toxicity can cause kidney disease too and ultimately renal failure. Since cadmium is responsible for osteoporosis (skeletal damage) been discovered in animal and human studies, it can cause bone mineralisation. Exposure to cadmium can also cause disruptions in the metabolism of calcium, renal stone formation and hypercalciuria. Cadmium is also listed by the International Organization for Research on Cancer as a Category 1 human carcinogen. The primary cause of cadmium absorption in smoking is cigarettes, so smokers are more vulnerable than non-smokers to cadmium intoxication (Altunatmaz *et al.*, 2018).

Chromium

The hexavalent form of chromium possesses the most toxic traits thus increasing its harming potential. On the other hand, the other species like chromium (III) compounds are much less toxic with countable to no health issues sometimes. Chromium (VI) is seemingly corrosive and also induces allergic reactions. The chances of inflammation in nose epithelial cells and even ulcers may also be caused by increased chromium (VI) amounts when exposed by inhalation. Sperm damages in male reproductive system, anaemia, inflammation in stomach and small intestine are also sighted in many cases. In case humans are exposed to extremely high doses of chromium (VI) compounds, serious cardiovascular, respiratory, haematological, gastrointestinal, hepatic, renal problems are seen in individuals. Human sensitivity to higher concentrations of chromium compounds may contribute to inhibition of erythrocyte glutathione reductase, which in turn limits the capacity to convert haemoglobin to methemoglobin. The experiments carried out *in vivo* and *in vitro* have successfully shown that chromium through different ways is capable of doing damage to DNA and can contribute to the development of chromosomal aberrations, DNA adducts, changes in chromatid sister replication, and transcription of DNA. It has bee also evidenced that chromium is responsible for fostering human carcinogenicity with an increase in number of tumours in the stomach.

Iron

Iron and its derivative salts, including sulphates, sulphate heptahydrates or sulphate monohydrates, have toxicity to lower levels when skin or respiratory systems are the routes of exposure. It brings toxicity in biological system. Iron toxicity occurs in four levels. Gastrointestinal symptoms, such as vomiting, diarrhoea, are characteristic of the first stage starting 6 h after iron overdose. The latter stages last for approximately 6 weeks which proceeds from symptoms like lethargy, metabolic acidosis, hypotension, hepatic necrosis, tachycardia, shocks and ultimately leading to death. Usually, the fourth and final stage of iron overdose happens within 2–6 weeks. This stage is remarkably identified by the developed strictures and by gastrointestinal inflammation. In cases where workers are exposed to asbestos for longer durations, 30% iron present in asbestos may cause induction of lung cancer. This can be linked in a way that as iron has been known to create free radicals resulting DNA damage by DNA oxidation that has been influencing the initiation of cancer. As the high amount of iron present in meat, the countries where meat eating population is high, more people are at the risk of cancer due to the fact that excess iron absorption increases the risk of cancer (Sanchez-Bayo and Goka, 2016; Altunatmaz *et al.*, 2018).

Insecticides

To regulate their body temperature, honeybees drink water while hunting for nectar. The remnants of agrochemicals in soil move into creeks and ponds after irrigation with the help of water stream. They thus make toxic the open water source for honeybees. The hymenopteran insects prefer to drink freshwater from lakes, ponds, river, small irrigation ditches, but what if all these sources are polluted with residues of toxic pesticides. The honeybees which drink this contaminated water accumulate these toxic substances in their hive.

Insecticides are highly toxic for honeybees while herbicides are comparatively nontoxic. Beekeeper should not establish their apiculture farm near the large commercial farming areas where pesticides are extensively applied.

The highly water-soluble chemicals such as fipronil and neonicotinoids are more poisonous and do not degrade easily in environment. Due to their high solubility in water, systemic pesticides move into rivers from the large agriculture areas and from there drain into remote areas where wild bees also get affected. Farmers use pesticides and fertilisers in seed dressing. The residues of these chemicals are sucked up by weeds and other flowering crops and ultimately they accumulate in nectar of those flowers which grow on treated soil. The chemicals used in agriculture affect the honeybees not only by their toxic nature but also by their specific mode of action. Neonicotinoids compounds act as a slow poison at low doses and excluding the many detrimental effects they cause. They have potential to destroy whole colony of honeybees if consumed for longer periods continuously. Many studies found immune suppressive effect exerted by neonicotinoids and fipronil which made honeybees more prone to parasitic infections such as *Nosema* and viral diseases which are transmitted by *Varroa* mites. The neurotoxic insecticides ingested by honeybees through water and food ultimately cause many types of diseases which cost a huge loss to overall ecosystem (Al-Waili *et al.*, 2012).

Long-term exposure to pesticides by any indirect method causes many health problems such as genetic alteration, circulatory disorder, digestive illness, cancer, and if appropriate steps are not taken in time the results are fatal. The ministry of agriculture in India identified aldrin, mirex, toxaphene, chlordane, hexachlorobenzene,

DDT, heptachlor, endrin and dihedron as persistent organic pollutants. There is evidence that these persistent organic pollutants affect the normal function of reproductive system of human beings (Rexilius, 1986).

Fungicides

Honey can be contaminated by fungicides used against harmful fungal pathogens in cultivated crops (Szczęsna *et al.*, 2011). The crops of economic importance which are very helpful in production of honey also are protected from pathogens by using those fungicides which have been approved by the responsible authorities as nontoxic to honeybees. For example, an exceptionally rich source of honey production is the intensive cultivation of oilseed rape in Schleswig-Holstein (FRG) (Rüegg, 1995). The use of insecticides and fungicides, which are legally licensed and confirmed to be non-toxic to honeybees, is only allowed to prevent crop losses of economic magnitude due to pests. A study in 1984 assessed samples of rape honey and analysed by gas-liquid chromatography for confirmation of the fungicides procymidone and vinclozoline. The recorded minimum values from the results revealed that pest control measures during crop cultivation inevitably contribute well determinable residues of the applied compounds in honey thus imposing negative effect on human health (Rüegg, 1995).

In countries like Poland and Switzerland, residues of the fungicides vinclozolin, penconazole, iprodione, captan, dithianon, pyrifenox, cyproconazole, methyl thiophanate and difenoconazole which were applied during the cultivation of crops were detected from the extracted honey considered for investigations (Büchler and Volkmann, 2003; Velicer *et al.*, 2004).

Antibiotics

Antibiotic residues have become a serious consumer concern in terms of honey consumption. Some antibiotics used commercially during apicultural practices are capable of producing specific toxic reactions in a sizeable population taking honey as a dietary component, while some others are capable of producing allergic or hypersensitivity reactions. For example, lactam antibiotics when it is taken at very low doses symptoms like cutaneous rashes, dermatitis, gastrointestinal symptoms and anaphylaxis usually develop (Paige *et al.*, 1997; Vass *et al.*, 2008).

Microbiologic risks, carcinogenicity, fertility consequences and teratogenicity are the long-term effects of exposure to antibiotic residues. One of the main health issues of human beings is microbiological impact. Any medications can induce cancer in human beings, such as nitrofurans and nitroimidazole. Similarly, at very low doses, certain medications can cause reproductive and teratogenic effects (Roe, 1985; Barganska *et al.*, 2011).

All these long-term effects of exposure to antibiotic residues also found in honey. Some medications possess capability of cancer induction in human beings, such as nitrofurans and nitroimidazole (Hubbard and Fidanze, 2007). Likewise, at very low doses, certain medications can cause reproductive and teratogenic effects.

A very serious consequence is that metals, pesticides, chemicals, antibiotics residues which are consumed along with honey bring resistance in bacterial populations.

The Centers for Disease Control and Prevention and World Health Organization (WHO) has raised a troubling concern that due to their use and overuse as antibiotics in apiculture, many bacterial pathogens are becoming resistant to the most widely used antibiotic treatments (Borm *et al.*, 2006; Chambers *et al.*, 2019).

Hydroxymethylfurfural

For maintenance of freshness of honey, to maintain its quality and increase the shelf life, techniques like heating or sterilisation are implemented. A concern connected with these sterilisation methods is that heating leads to formation of non-compositional compounds in honey which are dangerous to human health. The high toxic potential and widespread occurrence of HMF has attracted the attention in terms of toxicity issues related to honey (van Putten *et al.*, 2013). Many underlying factors are primarily responsible for the formation of HMF in honey during storage like metallic containers, humidity and the thermal and photochemical stresses. There is drastic increase in HMF concentrations when temperature or storage time is increased. In the case of varying pH, pyrolytic and dry conditions, HMF formation may take place via alternative pathways. Instances of varying HMF concentrations in honey from different sources have also been reported. Results from sample-based analysis suggest that honey was responsible for certain amounts of HMF in humans leading to health-related complications. To date, no mitigation techniques explicitly intended to minimise the concentration of HMF in honey have been identified, which is a concern due to the presence of HMF and its precursors as well as its development pathways. High HMF concentrations have also been shown to cause cytotoxic, irritating symptoms and even carcinogenic activity. The studies carried out for carcinogenic activity of HMF revealed the formation of aberrant crypt foci (preneoplastic lesions). HMF has been found possessing mutagenic potential as well. Both of these observations show that in rats, HMF is carcinogenic (Choudhary *et al.*, 2020). However, in human beings, the effects of 5-HMF found in rodents are not predicted. Therefore, the possible carcinogenic effects of 5-HMF on humans are a healthy area for future research to study. The key concern with respect to HMF is its conversion to SMF, its nephrotoxic capacity, its introduction of moderate liver damage and its increased extensive damage to the kidneys, particularly the proximal tubules. This suggests that the risk associated with HMF exposure from honey to humans is very high and can lead to detrimental effects on health.

Phytotoxins

The probability of transfer of plant toxins in the honey produced from their nectar is also responsible for honey toxicity. The toxicity is due to many secondary metabolites present in plants such as grayanotoxins (GTX), hyoscyamine, pyrrolizidine alkaloids, saponin, hyoscine, tutin, gelsemine and oleandrigenin. Honey collected from plants included in families like Ericaceae (*Andromeda* sp., *Rhododendron ponticum*, *Kalmia* sp., *Lleucothoe* sp., *Lyonia* sp., *Pieris* sp.) has been reported toxic. When bees collect nectar from these flowers, it has been reported that the resulting honey induces symptoms of toxicity in humans. GTXs are found as a constitutive component of honey (mad honey) when it is produced from the nectar of *R. ponticum* plant (Kurtoglu *et al.*, 2014). The honey thus produced is found toxic to humans only and not the bees. It has been reported that mad honey intoxication leads to cardiac conduction abnormalities and sometimes death. Many deaths have been reported in countries, including Australia and New Zealand. GTXs are remarkably identified by initial symptoms of intoxication such as salivation, vomiting, dizziness, weakness and paresthesia. When the level reaches higher doses symptoms like weakness in muscles, coordination loss, elevation in ST segment and electrocardiographic changes of bundle branch block are usually seen.

The nectar from *Andromeda* flowers even paralyses the limbs and diaphragm causing death. GTXs generally bind to specific sites in the sodium (Na^{2+}) channels and modify their function by producing hyperpolarisation in transmembrane potential. Sickness or even death can be caused by honey from the *Kalmia latifolia* flowers, the calico bush found in the northern region of the United States, and other plants such as 'wharangi bush' *Melicope ternata*. The inhabitants around the Yucatán Peninsula from Pre-Columbian period were aware of hazards of poisonous honey, although the production of honey was done by non-native stingless bees and not by honeybees. Depending on the type of toxin, the effects of honey poisoning can vary, but they typically include vomiting, palpitations, nausea, dizziness, convulsions, headache and even death (Islam *et al.*, 2014).

Conclusion

The toxicity of honey due to contamination through various routes has emerged as a serious concern with potential health hazards in its wake. Major causes of contamination are through improper apicultural practices. With the technology advancement the identification of contaminants present in honey has become easier. Toxic substances in honey are responsible for causing acute and chronic health affects more likely among children and the elderly. In general, the concentration of toxicants in honey determines whether a substance will pose hazard to the health status of individual or not because limited detoxification capacity. As a rule of thumb, the apiculturists should follow the conventional ways of apiculture that are known to be safe and in which a balanced and varied honey composition is maintained so that exposures to certain types of toxic substances can be kept limited.

REFERENCES

Altunatmaz, S. S., Tarhan, D., Aksu, F., Ozsobaci, N. P., Or, M. E., & Barutcu, U. B. 2018. Levels of chromium, copper, iron, magnesium, manganese, selenium, zinc, cadmium, lead and aluminium of honey varieties produced in Turkey. *Food Science and Technology (AHEAD)* 39:392–397.

Al-Waili, N., Salom, K., Al-Ghamdi, A., & Ansari, M. J. 2012. Antibiotic, pesticide, and microbial contaminants of honey: human health hazards. *The Scientific World Journal* 2012:1–9.

Bargańska, Ż., Namieśnik, J., & Ślebioda, M. 2011. Determination of antibiotic residues in honey. *TrAC Trends in Analytical Chemistry* 30(7):1035–1041.

Bartha, S., Taut, I., Goji, G., Vlad, I. A., & Dinulică, F. 2020. Heavy metal content in polyfloral honey and potential health risk. A case study of Copşa Mică, Romania. *International Journal of Environmental Research and Public Health* 17(5):1507.

Borm, A. A., Fox, L. K., Leslie, K. E., Hogan, J. S., Andrew, S. M., Moyes, K. M., ... & Norman, C. 2006. Effects of prepartum intramammary antibiotic therapy on udder health, milk production, and reproductive performance in dairy heifers. *Journal of Dairy Science* 89(6):2090–2098.

Büchler, R., Volkmann, B. 2003. Rückstände von Carbendazim und anderen Fungizidenim Bienenhonigaufgrund der Blütespritzung von Winterraps. *Gesun de Pflanzen* 55:217–221.

Chambers, A., MacFarlane, S., Zvonar, R., Evans, G., Moore, J. E., Langford, B. J., ... & Garber, G. 2019. A recipe for antimicrobial stewardship success: using intervention mapping to develop a program to reduce antibiotic overuse in long-term care. *Infection Control & Hospital Epidemiology* 40(1):24–31.

Choudhary, A., Kumar, V., Kumar, S., Majid, I., Aggarwal, P., & Suri, S. 2020. 5-Hydroxymethylfurfural (HMF) formation, occurrence and potential health concerns: recent developments. *Toxin Reviews*, 2020:1–17.

Grigoryan, K. Safety of honey. 2016. In Regulating safety of traditional and ethnic foods 2016 Jan 1 (pp. 217–246). Academic Press

Hubbard, R. D., & Fidanze, S. 2007. Therapeutic areas II: cancer, infectious diseases, inflammation and immunology and dermatology. *Comprehensive Medicinal Chemistry II*.

Islam, M. N., Khalil, M. I., Islam, M. A., Gan, S. H. 2014. Toxic compounds in honey. *Journal of Applied Toxicology* 34(7):733–742.

Jansen, S. A., Kleerekooper, I., Hofman, Z. L., Kappen, I. F., Stary-Weinzinger, A., van der Heyden, M. A. 2012. Grayanotoxin poisoning: 'mad honey disease' and beyond. *Cardiovascular toxicology* 12(3):208–215.

Kılıç Altun, S., Dinç, H., Paksoy, N., Temamoğulları, F. K., & Savrunlu, M. 2017. Analyses of mineral content and heavy metal of honey samples from south and east region of Turkey by using ICP-MS. *International Journal of Analytical Chemistry* 2017:1–6.

Kurtoglu, A. B., Yavuz, R., & Evrendilek, G. A. (2014). Characterisation and fate of grayanatoxins in mad honey produced from *Rhododendron ponticum* nectar. *Food Chemistry*, 161:47–52.

Matin, G., Kargar, N., & Buyukisik, H. B. 2016. Bio-monitoring of cadmium, lead, arsenic and mercury in industrial districts of Izmir, Turkey by using honey bees, propolis and pine tree leaves. *Ecological Engineering* 90:331–335.

Paige, J., Tollefson, L., Miller, M. 1997. Public health impact on drug residues in animal tissues. *Veterinary and Human Toxicology* 9:1–27.

Rexilius, L. 1986. Residues of pesticides in oilseed-rape honey (crop 1984) from Schleswig-Holstein-a study of the actual state. *Nachrichtenblatt des Deutschen Pflanzenschutzdienstes (Germany, FR)*.

Roe, F. J. 1985. Safety of nitroimidazoles. *Scandinavian Journal of Infectious Diseases. Supplementum* 46:72–81.

Rüegg, J. 1995. Monilia im Obstbau—Prüfung von Fungiziden in der biologischen ind Integrierten Produktion. *Obst- und Weinbau* 131:228–230.

Sall, M. L., Diaw, A. K. D., Gningue-Sall, D., Efremova Aaron, S., & Aaron, J. J. 2020. Toxic heavy metals: impact on the environment and human health, and treatment with conducting organic polymers, a review. *Environmental Science and Pollution Research* 27:29927–29942.

Sanchez-Bayo, F., & Goka, K. 2016. Impacts of pesticides on honey bees. *Beekeeping and Bee Conservation-Advances in Research*, 4:77–97.

Solayman, M., Islam, M. A., Paul, S., Ali, Y., Khalil, M. I., Alam, N., Gan, S. H. 2016. Physicochemical properties, minerals, trace elements, and heavy metals in honey of different origins: a comprehensive review. *Comprehensive Reviews in Food Science and Food Safety* 15(1):219–233.

Szczęsna, T., Rybak-Chmielewska, H., Waś, E., Kachaniuk, K., & Teper, D. 2011. Characteristics of Polish unifloral honeys. I. Rape honey (*Brassica napus* L. Var. Oleifera Metzger). *Journal of Apiculture Science* 55(1):111–119.

Ullah, S., Khan, S. U., Saleh, T. A., Fahad, S. 2018. Mad honey: uses, intoxicating/ poisoning effects, diagnosis, and treatment. *RSC Advances* 8(33):18635–18646.

van Putten, R. J., Van Der Waal, J. C., De Jong, E. D., Rasrendra, C. B., Heeres, H. J., & de Vries, J. G. (2013). Hydroxymethylfurfural, a versatile platform chemical made from renewable resources. *Chemical Reviews*, 113(3):1499–1597.

Vass, M., Hruska, K., & Franek, M. 2008. Nitrofuran antibiotics: a review on the application, prohibition and residual analysis. *Veterinarnimedicina* 53(9):469–500.

Velicer, C. M., Heckbert, S. R., Lampe, J. W., Potter, J. D., Robertson, C. A., Taplin, S. H. 2004. Antibiotic use in relation to the risk of breast cancer. *The Journal of the American Medical Association* 291(7):827–835.

8

Current Status and Future Strategies to Increase Honey Production in India

Swati Jamwal
Himachal Pradesh University, Shimla, India

Neha Sharma and Anjli Dhiman
Shoolini University, Solan, India

Shailja Kumari
Career Point University, Hamirpur, India

CONTENTS

DOI: 10.1201/9781003175964-8

Introduction

Honey is a viscous food substance made from flower nectar by Honey Bees (Dashora *et al.*, 2011). The presence of glucose and sucrose (monosaccharides) gives it a sweet taste (White, 1975, 2008). Small amounts of minerals, vitamins, proteins and amino acids are also present in honey, in addition to 13–23 percent water (White and Doner, 1980; Harrill, 1998; Downey *et al.*, 2005; Lewkowski *et al.*, 2019).

The minerals present in honey include calcium, copper, iron, magnesium, manganese, potassium, sodium, chlorides, phosphates, silicates and sulphates (Ajibola *et al.*, 2012). Mineral concentrations vary in honeys and mineral-rich honeys are dark in colour (Ajibola *et al.*, 2012). Some of the vitamins present in honey include ascorbic acid, pantothenic acid, niacin and riboflavin (Ajibola *et al.*, 2012).

Honey also contains other constituents like pollen, hydroxymethylfurfural (HMF) and enzymes (Da Silva *et al.*, 2016; Lewkowski *et al.*, 2019). Traces of pollen are also present in honey. HMF is formed naturally in honey over a period of storage (Shapla *et al.*, 2018). The enzymes mainly present in honey include invertase (sucrase), diastase (amylase), glucose oxidase, catalase and acid phosphatase (Seeley, 1978; Afroz *et al.*, 2016). These enzymes come from the bees during process of honey ripening or from the plant where the bee foraged (Afroz *et al.*, 2016).

Monosaccharides, mineral, vitamins, proteins and enzymes make it a very useful/ valuable product for humans. It's been used by humans because of its nutritional and medicinal properties for ages, nearly about 5500 years (Adebolu, 2005; Ashrafi *et al.*, 2005). Traditionally, honey was considered as a medicine (Çelik and Aşgun, 2020). Honey has anti-inflammatory, antibacterial, antidiabetic and antioxidant properties and it is also used to treat respiratory, gastrointestinal, cardiovascular and nervous system diseases (Samarghandian *et al.*, 2017).

These days honey is being recognised healing agent when taken orally or applied on wounds and burns (McLoone *et al.*, 2016). The ascorbic acid in honey provides relief in sore throat by inhibiting growth *Streptococcus pyogenes* (Salh, 2019). Honey has also shown antibacterial property against *Helicobacter pylori* which is the main cause of ulcers (Molan, 2001).

Apart from the medicinal value of honey, it also adds nutritional value to human diet (Maughan, 2002). Honey has been used as sweetener for ages and still a widely used source of sugars (Ramya and Anitha, 2020). It is also used to make honey wines, beers, breakfast cereals, bakery goods and value-added products (Orina, 2014).

Honey can be of different types on the basis of its origin and industrial use as discussed in the following sections.

Honey Categories on the Basis of its Origin

Blossom Honey

Honey mainly obtained from the nectar of flowers is called blossom honey (Pita-Calvo *et al.*, 2016).

Honeydew Honey

Honeydew honey is produced by honey bees of Rhynchota genus as they collect the plant sap and secrete it back (Agbajor *et al.*, 2017).

Monofloral Honey

Honey made up of nectar collected from same type flowers is called mono floral honey (Alvarez-Suarez *et al.*, 2014).

Multifloral Honey

Multifloral honey is made of by mixing nectar from different types of flowers (Agbajor *et al.*, 2017).

Honey Categories on the Basis of use or Trade

Table Honey

Table honey has very less antimicrobial activity and used for consumption directly or as a sweetener in beverages (Cooper and Jenkins, 2009).

Industrial Honey

Industrial or baker's honey is a lower quality because the presence of high concentration of HMF due to overheating (Bogdanov and Martin, 2002). Industrial honey is good for the bakery and pharmaceutical industries (Orina, 2014).

Current Status of Honey Production in India

India has arable land which covers area of 159.7 million hectare and bee foraging crops. Both of these factors contribute to the potential of the country to harbour about 200 million bee colonies. In the past 10 years, a steady increase in honey production

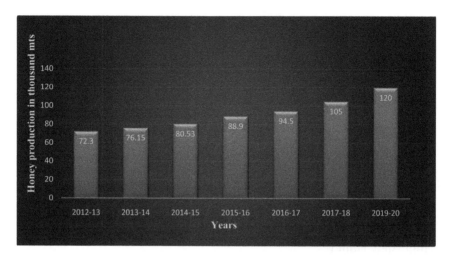

FIGURE 8.1   Honey production in India from 2012 to 2020 as per NBB data.

has been observed as a result of continuous efforts though different programmes like National Beekeeping & Honey Mission (NBHM). Approximately 200% increase in honey production has been seen in the last decade.

As per the data on National Bee Board (NBB) and Ministry of Agriculture and Farmers Welfare, steady increase in India's honey production has been seen. Honey production increased to 1,20,000 metric tonnes (MTs) as compared to the 76,150 MTs in 2013–2014, which is 57.58% increase (Figure 8.1). Honey export rate has also increased in India by 109.80% (Figure 8.2) as export of honey has increased from 28,378.42 MTs (2013–2014) to 59,536.74 MTs (2019–2020).

The beekeeping sector in India strengthens yearly. In total, 28 states of India are involved in beekeeping (Table 8.1). In 2019, 1,559,771 bee colonies, 9580 individual

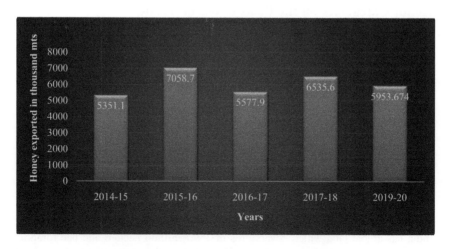

FIGURE 8.2   Honey exported from India from 2014 to 2020 as per NBB data.

TABLE 8.1

State wise Details of Beekeepers Registered with the NBB up to 31 December 2019 (Source: NBB)

Sr. No.	Name of State	Individual Beekeepers Entities Registered	Beekeeping and Honey Societies Entities Registered	Companies Entities Registered	Firms Entities Registered	Self Help Group Entities Registered	Total Entities Registered
1	Andhra Pradesh	31	1	–	–	–	32
2	Assam	126	–	1	–	–	127
3	Bihar	859	1	–	–	–	860
4	Chhattisgarh	1	1	–	–	–	2
5	Delhi	38	2	7	2	–	49
6	Goa	4	–	–	–	–	4
7	Gujarat	77	–	3	1	–	81
8	Haryana	906	2	3	6	–	917
9	Himachal Pradesh	361	2	–	–	–	363
10	Jammu and Kashmir	226	1	–	–	–	227
11	Jharkhand	55	–	1	–	–	56
12	Kerala	13	2	–	–	–	15
13	Madhya Pradesh	269	2	–	1	–	272
14	Maharashtra	63	1	1	–	1	66
15	Nagaland	232	–	–	–	–	232
16	Odisha	30	–	–	–	–	30
17	Punjab	1156	–	4	9	–	1169
18	Rajasthan	587	3	–	8	–	598
19	Tamil Nadu	7	–	–	1	–	8
20	Uttarakhand	434	2	1	2	–	439
21	Uttar Pradesh	2699	27	6	10	2	2744
22	West Bengal	673	–	–	–	–	673
23	Karnataka	658	–	–	–	–	658
24	Manipur	2	1	–	–	–	3
25	Tripura	2	–	–	–	–	2
26	Arunachal Pradesh	63	–	–	–	–	63
27	Mizoram	6	–	–	–	–	6
28	Pondicherry	2	–	–	–	–	2
Total		9580	48	27	40	3	9698

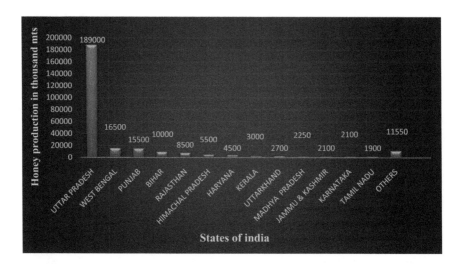

FIGURE 8.3   Honey production by state in India during 2017–2018 (source NBB).

beekeepers, 48 beekeeping and honey societies, 27 companies, 40 firms and 3 self-help groups are involved in apiary business and were registered with NBB (Table 8.1). On 11 February 2021, the total number of registered honeybee colonies rose to 16.00 lakhs and individual beekeepers, honey societies/firms/companies registered with NBB to 10,000 (https://pib.gov.in/PressReleaseDetailm.aspx?PRID=1697113).

Top Honey Exporter States of India

As per data from NBB among 28 states of India involved in honey production, the top states in honey production are Uttar Pradesh, West Bengal, Punjab and Bihar (Figure 8.3). More than 50% of the total honey produced in the country comes from these four states. Three other states also produce considerable amount of honey are Rajasthan, Himachal Pradesh and Haryana. The Southern states of the country contribute approximately 25% of total honey production.

Types of Honey Produced and Exported from India

Honey can be mono or multifloral. Monofloral or unifloral honeys have predominance of nectar collected from a single type of plant. Multifloral or wildflower honeys are formed from nectar collected from different varieties of flowers. The following types of honeys are produced in India (http://msmedikanpur.gov.in):

1. Rapeseed/mustard honey
2. Eucalyptus honey
3. Lychee honey

4. Sunflower honey
5. Karanj/Pongamia honey
6. Multiflora Himalayan honey
7. Acacia honey
8. Wild flora honey
9. Multi-monofloral honey

Current Strategies to Increase Honey Production and Export

In India, until 1990 honey production and beekeeping were an unexplored sector. In 1993, the National Beekeeping Development Board was constituted. Later in 2000, NBB was registered by Small Farmers Agribusiness Consortium (SFAC) under Society Act, 1860. NBB was reconstituted in 2006 with Secretary (A and C) as Chairman. National Horticulture Mission (NHM) and Horticulture Mission for North East and Himalayan States (HMNEH) Schemes in Beekeeping were implemented by NBB in 2014 to develop beekeeping sector.

National Beekeeping and Honey Mission

NBHM focused on the overall promotion and development of scientific beekeeping in the country to bring 'Sweet Revolution'. NBHM is being implemented through NBB. Amount of Rs. 500 crores was allocated by Indian government for National Beekeeping & Honey Mission (NBHM) for three years (2020–2021 to 2022–2023). The mission was announced as part of the Atma Nirbhar Bharat scheme.

The main objective of NBHM is to promote holistic growth of beekeeping industry. Under this mission, Integrated Beekeeping Development Centre (IBDC)s/CoE, honey testing labs, bee disease diagnostic labs, custom hiring centres, apitherapy centres, nucleus stock, bee breeders etc. will be established. This scheme has following three mini-missions:

1. Mini-Mission-I: aims to create awareness about scientific beekeeping.
2. Mini-Mission-II: targets the management of beekeeping after harvesting and deals with collection, processing, storage, marketing and value addition of beehive products.
3. Mini-Mission-III: focused on Research and advancement of technology in beekeeping sector.

Under NBHM Scheme 11, projects of Rs. 2560 lakhs were sanctioned to bring awareness about scientific breeding and capacity building, technology demonstrations on impact of Honey Bees on yield enhancement etc. It also targets to keep farmers informed about the distribution of specialised beekeeping equipment for producing high-value products, viz. Royal Jelly, Bee Venom, Comb Honey, etc. These schemes also focus on studies on potential of high-altitude honey, on special honey production in Kannauj and Hathras districts of Uttar Pradesh, and treatment of colon cancer using mustered during 2020–2021.

Two world-class 'State of the Art Honey Testing Labs' were established under NBHM, one lab at National Dairy Development Board's (NDDB), Gujarat and other at Indian Institute of Horticultural Research (IIHR), Bengaluru. Lab is functional to test honey samples as per parameters laid by FSSAI.

Under NBHM, five Farmer Producer Organizations (FPOs) of beekeepers were also launched in November, 2020. These FOPs are constituted in states of Uttar Pradesh, West Bengal, Bihar, Madhya Pradesh and Rajasthan.

Integrated Beekeeping Development Centres (IBDCs) as role model of beekeeping were established in 16 States, one in each state. Total 16 IBDCs are in the states of Jammu and Kashmir, Himachal Pradesh, Punjab, Haryana, Delhi, Uttar Pradesh, Bihar, Uttarakhand, Manipur, Madhya Pradesh, Tripura, Andhra Pradesh, West Bengal, Arunachal Pradesh, Tamil Nadu and Karnataka.

Ministries, Departments, Agencies and Institutions dealing with Beekeeping in India

Different organisations are working together to enrich and improve the beekeeping and honey production sector in India. The following listed ministries, departments and institutes are involved in beekeeping sector and have their important role directly or indirectly to increase honey production in country:

1. Agricultural and Processed Food Products Export Development Authority (APEDA)
2. All India Coordinated Research Project on Honey Bees and Pollinators (AICRP-HB&P)
3. Central Beekeeping Research & Training Institute (CBRTI)
4. Department of Agriculture Cooperation (DAC)
5. Farm Sector Development Funds (FSDF)
6. Food Safety Standard Authority of India (FSSAI)
7. Honeybee Research Institute (HBRI)
8. Integrated Beekeeping Development Centres (IBDCs)
9. Khadi and Village Industries Commission (KVIC)
10. Ministry of Agriculture and Farmers Welfare (MoA & FW)
11. Ministry of Agriculture and Farmers Welfare (MoA & FW)
12. Ministry of Commerce & Industry (MoC & I)
13. Ministry of Finance (MoF)
14. Ministry of Health and Family Welfare (MoH & FW)
15. Ministry of Micro, Small and Medium Enterprises (MoMSME)
16. National Bank for Agriculture and Rural Development (NABARD)
17. NBB
18. National Horticultural Mission (NHM)
19. State Agricultural Universities (SAUs)

Above-mentioned ministries, departments, agencies, institutions and other than these play their role to carryout research, to provide financial assistance and training to improve and increase the honey production. They also set the quality standards of honey to increase export. Following roles are played by these ministries, departments, agencies, institutions.

Research in Area of Beekeeping to Increase Quality and Quantity of Honey

The Indian Council of Agricultural Research (ICAR), Central Agricultural Universities (CAUs) and State Agricultural Universities (SAUs) in Dept of Agricultural Research and Education (DARE) under Ministry of Agriculture and Farmers Welfare (MoA & FW) assist the Government of India by doing research in field of beekeeping and its role in cross-pollination and increase in crop productivity.

Financial Support in Development in Beekeeping and Honey Production Sector

Both MoA & FW and MoF control Development in beekeeping and honey production sector. The NBB and National Horticultural Mission (NHM) of Department of Agriculture Cooperation (DAC) under MoA & FW provide financially supports to farmers through Integrated Beekeeping Development Centres (IBDCs) in State Agricultural Universities and Private beekeeping industry. NABARD under MoF, through its Farm Sector Development Funds (FSDF), provides funds through state governments and NGOs to support beekeeping.

Training to Increase Honey Production

The Khadi and Village Industries Commission (KVIC) in the Ministry of MSME is focused on development of beekeeping sector mainly for increased honey production through Central Beekeeping Research & Training Institute (CBRTI). These provide training to beekeepers.

Quality Control of Honey

The Food Safety Standard Authority of India (FSSAI) under Ministry of Health and Family Welfare sets the quality standards for honey and as well as for bee products and assures their implementation.

Export of Honey

Export of honey is controlled by Agricultural and Processed Food Products Export Development Authority (APEDA) under Ministry of Commerce & Industry. The Export Inspection Council (EIC) is the official export certification body of India which ensures the quality and safe export of honey products from India.

Future Strategies to Increase Honey Production

A steady increase in honeybee colonies and honey produced in India has increased considerably in the last decade. Though more efforts are required to increase honey production in India, as per NBB India has potential to harbour 200 million bee colonies. Improvement in and more effective implementation of scientific beekeeping will be useful to utilise maximum potential of India in honey production. The following listed strategies can result in increased honey production in India.

Pest Management

Pests of honeybee include insects, birds, etc. (reference) possess harm to bee colonies. The following methods and techniques can be useful in pest management.

Genetic Resistance Honeybees

Genetically resistant honeybee species should be used preferably in beekeeping sector as they can combat disease outbreaks efficiently (Lapidge *et al.*, 2002; Goode *et al.*, 2005; Evans and Spivak, 2010). Genetically resistant line honey bees can be achieved by improving the ability of honeybees to detect and destroy infected broods and improving their hygiene behaviour (Aronstein and Murray, 2010).

Cultural Practices

Infected combs should be removed by beekeeper time to time. Reduction stress by providing adequate ventilating and humidity, feeding sugar syrup or sugar water, time-to-time cleansing and changing of hive combs in every two years can be beneficial to control spread of pest (Flores *et al.*, 2005, Palacio *et al.*, 2010; Food and Agricultural Organization of the United Nations). Cultural practice of thorough destruction, possibly by burning the infected and old hives, can be useful in pest management (Forsgren *et al.*, 2018). Arthropod pests can be managed by keeping bottom broods at a distance from ground, removing adhered debris and placing insect traps around beehives (HPMSP, 2008).

Biological Agents

Biological agents have been proved the most the most promising in terms of safety and eco-friendly control of parasitic and pathogenic pests of Honey Bees. For example, *Metarhizium anisopliae* and *Beauveria bassiana* known as Entomopathogenic fungus have been proved as promising agent to control Varroa mites (Shaw *et al.*, 2002; García-Fernández *et al.*, 2008). *Chelifer cancroides* and mite like *Stratiolaelaps scimitus*; *Mesostigmata*: *Laelapidae* have controlled Varroa pest effectively (Read *et al.*, 2014; Rangel and Ward, 2018). *Bracon hebetor* and *Apanteles galleriae* parasitoids are potential agents to control wax moths (Kwadha *et al.*, 2017). More exploration needs to be done in this field to identify beneficial microorganisms to control pests of Honey Bees (Arbia and Babbay, 2011; Rangel and Ward, 2018).

Insect Sterilisation Techniques

Contaminated beekeeping equipment can be sterilised by exposing to gamma radiations generated by decay of Cobalt-60 source (Hornitzky and Willis, 1983). Gamma-irradiation at optimal can be used to sterilise bee wax and honey (Wooton *et al.*, 1985; Baggio *et al.*, 2005). But this facility is very limited due to shortage of accessible equipment. More Efforts should be done to make this technique easily accessible to beekeepers.

Microwave heating is very effective in complete removal of yeast from honey. Employment of techniques like membrane processing, microfiltration and ultrafiltration can be very beneficial to produce enzyme-enriched and microorganism free honey (Little *et al.*, 1987; Subramanian *et al.*, 2007).

Management of Anthropogenic Activities and Climatic Factors Affecting Honey Production

Management of Anthropogenic Activities

Increased anthropogenic activities adversity affect the foraging areas for honeybees (Godfrey, 2015). Increased farming and grazing activities have led to deforestation and environmental degradation which resulted in reduced water resources, loss of nesting and foraging area for Honey Bees (Kaale *et al.*, 2002; Yanda and Madulu, 2005; Makero and Kashaigili, 2016; Lehébel-Péron *et al.*, 2016). However, Kovács in 2013 and Vaughan *et al.* in 2015 reported that honey production increases in response to flowering crops. Thus, controlled anthropogenic activities and growing more flowering crops can influence honey production positively.

Management of Climatic Factors

Harsh climatic conditions like very high temperature or heavy or low rainfall can affect the production of honey. Efforts can be made to minimise the damage be climatic condition by following methods.

Providing Necessary Climate Change Information to Beekeepers

Beekeepers have very limited or little knowledge of climatic factors interfering with honey production (Cherotich *et al.*, 2012; Li *et al.*, 2013). By proving timely and relevant information of climate factor and adaptation, it equips beekeeper to deal with climatic variations. Knowledge of climatic change factors will help farmers in management of ecologically sound apiary site, mitigating the risk climate change and take advantage of it. Therefore necessary training and campaigning should be done to pass this knowledge to farmers.

Adoption of Adaptation Strategies by Beekeepers

Gbetibou (2009) and Malisa and Yanda (2015) reported that different strategies like diversification of bee species, change in sites of beehives, increasing or decreasing the number of beehives, using a specific type of beehive either modern or traditional, change in apiary techniques and in periods of beekeeping operations, shifting to honey hunting from beekeeping activity were effective for adaptation to climate change. Thus, the effective climate change adaptation measures should be employed by beekeeper to prevent loss of bee colonies and to increase honey production.

Market Oriented Honey Production

Increased honey demand in market can lead to the increased honey production. To increase market demand for honey, the following strategies can be employed.

Quality of Honey

Honey collected through traditional methods contains impurities. So advanced technologies like centrifugation techniques should be used for collection and processing of honey.

Honey quality should meet the standards laid by Agricultural and Processed Food Products Export Development Authority (APEDA) and markets such as European Union, Japan and the United States. It will prove beneficial to increase export of honey.

Increasing Diversity of Honey Products

Value addition to the basic product can create a wide range of diverse products. Product diversity offers a wide range of products to satisfy the needs of customer, while value addition will increase the price of each gram of honey. This will stabilise income even during off seasons and will provide employment for other sectors. Stability in income and employment will encourage more honey production.

Marketing and Export of Honey

Campaigning to spread awareness about the benefits and quality if Indian honey should be done, a brand equity for honey should be created by collective efforts of APEDA, Ministry of Commerce & Industry and the Government of India for better marketing and to increase sale at better price. The development of efficient export marketing network can balance production and export honey (Shilpashree *et al.*, 2017).

Recognition of Beekeeping as Industry

Honey industry has potential to be recognised as agro-industry or agri-horticulture industry or forest-based industry. If international standards of honey quality are met, honey industry can be a major foreign exchange earner. The use of modern and advanced technologies to collect store, process and packing can be beneficial to meet foreign standard of honey quality. Beekeeping, honey production and production of value-added products can generate employment to many people.

Honey is being used in India for ages due to its medicinal properties (Arawwawala and Hewageegana, 2018). Still, till 1990 potential of beekeeping and honey production was unexplored. In 1993, National Beekeeping Development Board was established. Later in the year 2000, 'National Bee Board' was constituted and was reconstituted in 2006. NBB is focused on overall growth of beekeeping sector by scientific beekeeping to increase crop productivity through pollination and honey production to benefit farmers financially. In 2020, the Government of India launched National Beekeeping and Honey Management to bring 'Sweat revolution' in country. The Government of India has allocated 500 crores for this mission. NBHM is focused on to increase beekeeping, honey production, honey export, crop productivity, farmer's income, advanced beekeeping technologies etc. As per NBB date, India has potential to harbour 200 million colonies of Honey Bees (https://nbb.gov.in). To exploit this potential, other schemes like NBHM should be launched. To increase honey production, genetically resistant species of honey bees should be preferred for beekeeping and climatic adaptation strategies must be practised by beekeepers Evans and Spivak, 2010; Malisa and Yanda, 2015).

REFERENCES

Adebolu, T.T., 2005. Effect of natural honey on local isolates of diarrhea-causing bacteria in southwestern Nigeria. *African Journal of Biotechnology*, *4*(10), pp. 1172–1174.

Afroz, R., Tanvir, E.M., Zheng, W. and Little, P.J., 2016. Molecular pharmacology of honey. *Clinical and Experimental Pharmacology*, *6*(3), pp. 1–13.

Agbajor, G. K., Otache, M.A. and Akpovona, A.E., 2017. Measurements and analysis of the electrical conductivity of selected honey samples in Nigeria. *Asian Journal of Physical and Chemical Sciences*, 3(3), pp. 1–5.

Ajibola, A., Chamunorwa, J. P. and Erlwanger, K.H., 2012. Nutraceutical values of natural honey and its contribution to human health and wealth. *Nutrition & Metabolism*, *9*(1), pp. 1–12.

Alvarez-Suarez, J.M., Gasparrini, M., Forbes-Hernández, T.Y., Mazzoni, L. and Giampieri, F., 2014. The composition and biological activity of honey: a focus on Manuka honey. *Foods*, *3*(3), pp. 420–432.

Arawwawala, L.D. and Hewageegana, H.G.S.P., 2018. Health benefits and traditional uses of honey: A review. *Journal of Apitherapy*, *2*(1), pp. 9–14.

Arbia, A. and Babbay, B., 2011. Management strategies of honeybee diseases. *Journal of Entomology*, *8*(1), pp. 1–15.

Aronstein, K.A. and Murray, K.D., 2010. Chalkbrood disease in honey bees. *Journal of Invertebrate Pathology*, *103*, pp. 20–29.

Ashrafi, S., Mastronikolas, S. and Wu, C.D., 2005. March. Use of honey in treatment of aphthous ulcers. *Abstract*, 1262, pp. 9–12.

Baggio, A., Gallina, A., Dainese, N., Manzinello, C. and Mutinelli, F., 2005. Gamma radiation: a sanitating treatment of AFB-contaminated beekeeping equipment. *Apiacta*, *40*, pp. 22–27.

Bogdanov, S. and Martin, P., 2002. Honey authenticity. *Mitteilungen aus Lebensmitteluntersuchung und Hygiene*, *93*(3), pp. 232–254.

Çelik, K. and Aşgun, H.F., 2020. *Apitherapy: Health and Healing from the Bees*. Tudás Alapítvány.

Cherotich, V.K., Saidu, O. and Bebe, B.O., 2012. Access to climate change information and support services by the vulnerable groups in semi-arid Kenya for adaptive capacity development. *African Crop Science Journal, 20*, pp. 169–180.

Cooper, R. and Jenkins, L., 2009. A comparison between medical grade honey and table honey. *Wounds, 21*, pp. 29–36.

Da Silva, P.M., Gauche, C., Gonzaga, L.V., Costa, A.C.O. and Fett, R., 2016. Honey: chemical composition, stability and authenticity. *Food Chemistry, 196*, pp. 309–323.

Dashora, N., Sodde, V., Bhagat, J., S Prabhu, K. and Lobo, R., 2011. Antitumor activity of *Dendrophthoe falcata* against Ehrlich ascites carcinoma in Swiss albino mice. *Pharmaceutical Crops, 2*, pp. 1–7.

Downey, G., Hussey, K., Kelly, J.D., Walshe, T.F. And Martin, P.G., 2005. Preliminary contribution to the characterisation of artisanal honey produced on the island of Ireland by palynological and physico-chemical data. *Food Chemistry, 91*(2), pp. 347–354.

Evans, J.D. and Spivak, M., 2010. Socialized medicine: individual and communal disease barriers in honey bees. *Journal of Invertebrate Pathology, 103*, pp. 62–72.

FAO (Food and Agricultural Organization of the United Nations), 2006. *Honeybee diseases and pests: A practical guide: Agricultural and food engineering technical report.*

Flores, J.M., Spivak, M. and Gutiérrez, I., 2005. Spores of *Ascosphaeraapis* contained in wax foundation can infect honeybee brood. *Veterinary Microbiology, 108*(1–2), pp. 141–144.

Forsgren, E., Locke, B., Sircoulomb, F. and Schäfer, M.O., 2018. Bacterial diseases in Honey Bees. *Current Clinical Microbiology Reports, 5*(4), pp. 18–25.

García-Fernández, P., Santiago-Álvarez, C. and Quesada- Moraga, E., 2008. Pathogenicity and thermal biology of mitosporic fungi as potential microbial control agents of *Varroa destructor* (Acari: Mesostigmata), an ectoparasitic mite of honey bee, *Apis mellifera* (Hymenoptera: Apidae). *Apidologie, 39*(6), pp. 662–673.

Gbetibou, G.A., 2009. Understanding farmers' perceptions and adaptations to climate change and variability: The case of the Limpopo Basin, South Africa. Environment and Production Technology Division. IFPRI (International Food Policy Research Institute), Discussion paper 00849. Washington, DC, USA.

Godfrey, G., 2015. Epidemiology of honey bee disease and pests in selected zones of Tigray region, northern Ethiopia, M.Sc. Thesis.

Goode, K., Huber, Z., Mesce, K.A. and Spivak, M., 2005. The relationships of honeybee (*Apis mellifera*) behaviors in the context of octopamine neuromodulation: Hygienic behavior is independent of sucrose responsiveness and foraging ontogeny. *Hormones and Behavior, 49*(3), pp. 391–397.

Harrill, R., 1998. Using a refractometer to test the quality of fruits and vegetables. *P. PUBLISHING, Éd.) Consulté le July, 20*, p. 2010.

Hornitzky, M.A.Z. and Willis, P.A., 1983. Gamma radiation inactivation of *Bacillus larvae* to control American foul brood. *Journal of Apicultural Research, 22*(3), pp. 196–199.

HPMSP (Honeybee pest management strategic plan), 2008. *Honey bee pest management strategic plan – The mid-Atlantic states in pest management strategic plan for Honey Bees in the mid-Atlantic states.* Southern Region IPM Center Virginia Tech North Carolina State University Maarec 2008.

Kaale, B.K., Ramadhani, H.K., Kimaryo, B.T., Maro, R.S. and Abdi, H., 2002. Participatory Forest Resource Assessment. Misitu Yetu Project, CARE Tanzania, Dar es Salaam, Tanzania.

Kovács, A., 2013. The role of beekeeping in production of oil crops. *Applied Studies in Agribusiness and Commerce*, 7(4–5), pp. 7–82.

Kwadha, C.A., Ongamo, G.O., Ndegwa, P.N., Raina, S.K. and Fombong, A.T., 2017. The biology and control of the greater wax moth. *Galleria Mellonella Insects*, 8(61), pp. 1–17.

Lapidge, K., Oldroyd, B. and Spivak, M., 2002. Seven suggestive quantitative trait loci influence hygienic behavior of Honey Bees. *Naturwissenschaften*, 89(12), pp. 565–568.

Lehébel-péron, A., Sidawy, P., Dounias, E. and Schatz, B., 2016. Attuning local and scientific knowledge in the context of global change: The case of heather honey production in southern France. *Journal Rural Studies*, 44, pp. 32–142.

Lewkowski, O., Mureşan, C.I., Dobritzsch, D., Fuszard, M. and Erler, S., 2019. The effect of diet on the composition and stability of proteins secreted by honey bees in honey. *Insects*, 10(9), pp. 282.

Li, H., Le, D., Elton, J.B. and Ian, N., 2013. Farmers' perceptions of climate variability and barriers to adaptation: Lessons learned from an Exploratory Study in Vietnam. *Mitigation and Adaptation Strategies for Global Change*. Netherlands: Springer.

Little, B., Gerchakov, L. and Udey, L., 1987. A method for sterilization of natural seawater. *Journal of Microbiological Methods*, 7(4–5), pp. 193–200.

Makero, J.S. and Kashaigili, J.J., 2016. Analysis of land-cover changes and anthropogenic activities in Itigi Thicket, Tanzania. *Advanced Remote Sensing*, 5, pp. 269–283.

Malisa, G.G. and Yanda, P.Z., 2015. Impact of climate variability and change on beekeeping productivity. The First Continental Symposium on Honey Production, Bee Health and Pollination Services in Africa, 6th September, 2015, Cairo, Egypt.

Maughan, R., 2002. The athlete's diet: nutritional goals and dietary strategies. *Proceedings of the nutrition Society*, 61(1), pp. 87–96.

McLoone, P., Oluwadun, A., Warnock, M. and Fyfe, L., 2016. Honey: a therapeutic agent for disorders of the skin. *Central Asian Journal of Global Health*, 5(1), pp. 241.

Molan, P.C., 2001. The potential of honey to promote oral wellness. *General Dentistry*, 49(6), pp. 584–590.

Orina, I.N., 2014. *Quality and safety characteristics of honey produced in different regions of Kenya* (Doctoral dissertation).

Palacio, M. A., Rodriguez, E., Goncalves, L., Bedascarrasbure, E. and Spivak, M., 2010. Hygienic behaviors of Honey Bees in response to brood experimentally pin-killed or infected with. *Ascosphaeraapis. Apidologie*, 41(6), pp. 1–11.

Pita-Calvo, C. and Vázquez, M., 2016. Differences between honeydew and blossom honeys: A review. *Trends in Food Science and Technology*, 59, pp. 79–87.

Ramya, H.N. and Anitha, S., 2020. Development of muffins from wheat flour and coconut flour using honey as a sweetener. *International Journal of Current Microbiology & Applied Science*, 9(7), pp. 2231–2240.

Rangel, J. and Ward, L., 2018. Evaluation of the predatory mite *Stratiolaelaps scimitus* for the biological control of the honey bee ectoparasitic mite *Varroa destructor*. *Journal of Apicultural Research*, 57(3), pp. 425–432.

Read, S., Howlett, B.G., Donovan, B.J., Nelson, W.R. and van Toor, R.F., 2014. Culturing chelifers (Pseudoscorpions) that consume Varroa mites. *Journal of Applied Entomology*, 138(4), pp. 260–266.

Salh, M.M.S., 2019. *Antimicrobial Activity of Aqueous Citrus limon Extract against Streptococcus pyogenes (Group A) Isolated from Sore Throat Patients in Jeddah City, Saudi Arabia* (Doctoral dissertation, Sudan University of Science & Technology).

Samarghandian, S., Farkhondeh, T. and Samini, F., 2017. Honey and health: A review of recent clinical research. *Pharmacognosy Research*, *9*(2), p. 121.

Seeley, T.D., 1978. Life history strategy of the honey bee, *Apis mellifera*. *Oecologia*, *32*(1), pp. 109–118.

Shapla, U.M., Solayman, M., Alam, N., Khalil, M.I. and Gan, S.H., 2018. 5-Hydroxymethylfurfural (HMF) levels in honey and other food products: effects on bees and human health. *Chemistry Central Journal*, *12*(1), pp. 1–18.

Shaw, K.E., Davidson, G., Clark, S.J., Ball, B.V., Pell, J.K., Chandler, D. and Sunderland, K.D., 2002. Laboratory bioassays to assess the pathogenicity of mitosporic fungi to *Varroa destructor* (Acari: Mesostigmata), an ectoparasitic mite of the honeybee, *Apis mellifera*. *Biological Control*, *24*(3), pp. 266–276.

Shilpashree, J., Serma Saravana Pandian, A. and Veena, N., 2017. Trade performance of natural honey in India - a Markov approach. *Bulletin of Environment, Pharmacology and Life Sciences*, *6*, pp. 111–114.

Subramanian, R., Hebbar, U.H. and Rastogi, N.K., 2007. Processing of honey. A review. *International Journal of food Properties*, *10*(1), pp. 127–143.

Vaughan, M., Hopwood, J., Lee-Mäder, E., Shepherd, M., Kremen, C., Stine, A. and Black, S.H., 2015. Farming for bees; Guidelines for Providing Native Bee Habitat on Farms. The Xerces Society for Invertebrate Conservation.

White, J.W.W., 1975. Composition of honey. In Crane, E. (Ed.), *Honey a comprehensive survey*. London: Heinemann, pp. 157–206.

White, J.W. and Doner, L.W., 1980. Honey composition and properties. *Agriculture Handbook*, *335*, pp. 82–91.

White, J.S., 2008. Straight talk about high-fructose corn syrup: what it is and what it ain't. *The American Journal of Clinical Nutrition*, *88*(6), pp. 1716–1721.

Wooton, M., Hornitzky, M.A.Z. and Beneke, M., 1985. The effects of gamma-radiation from cobalt-60 on quality parameters of Australian honey. *Journal Apicultural Research*, *24*(3), pp. 188–189.

Yanda, P.Z. and Madulu, N.F., 2005. Water resource management and biodiversity conservation in the Eastern Rift Valley Lakes, Northern Tanzania. *Physics & Chemistry of Earth*, *30*, pp. 717–725.

9

Honey and Honeybees as Potential Pollinators and Indicators of Environmental Pollution

Younis Ahmad Hajam
Career Point University, Hamirpur, India

Ankush Sharma
Sri Sai University, Palampur, India

Indu Kumari
Arni University, Kangra, India

Rajesh Kumar
Himachal Pradesh University, Shimla, India

CONTENTS

Introduction

Insects are wonderful creatures that contribute about 80% of the total animal population. They are capable of adjusting themselves to diverse types of climatic conditions. Insects are a very important part of the food chain and ecosystem, supporting human life directly or indirectly. Among insects, honeybees are the most interesting and beneficial ones, living in small colonies. They are highly civilised organisms that work well in the temperature range of 10–40°C. Honeybees are divided into three castes and are dedicated to given duties. They work selflessly and perform different duties throughout their life, which motivates humans. Honeybees are small insects that are sensitive to environmental changes also. Their efficiency to sense and respond to every change in the environment is incredible. Honeybees act as an excellent detector of environmental pollution, global warming and associated factors (Celli and Maccagnani, 2003). The existence of a honeybee is completely related to the environment in which it lives (Sadeghi *et al.*, 2012; Steen *et al.*, 2012). These insects have a very important role in the detection of environmental pollution in many ways: signalled by their high mortality rates due to mobile towers (Al-Akhras *et al.*, 2015)and concluded by the presence of residues of pesticides, fungicides, herbicides, heavy metals, radionuclide and other pollutants in bee products (Sadeghi *et al.*, 2012). The analysis of honey, wax and other products is often quite useful in assessing environmental quality. Honey produced from 'pollen', 'plant nectars' and 'honeydew' constitutes over 300 chemical components that come under diverse chemical compound groups.

Unknowingly, honeybees bring various pollutants from the environment, pollen and nectar and deposit these toxicants inside their hives. Therefore, plant protective items utilised in farming are not only the reason for the mass poisoning of honeybees, but also these get passed on to beehive products, particularly honey, affecting its quality, properties and causing a threat to the health of human beings (Kujawski and Namiesnik, 2008; Barganska *et al.*, 2013). In India, global warming, particularly during hot weather, has often been observed on the wild species of bees, *Apis dorsata* and *Apis florea*. Due to the partial melting of wax at long spells of elevated temperatures, honeybee combs become weaker and cannot withstand the heavy load and fall. Global warming also causes hurricanes, hot and gusty winds, other climatic changes and weather turmoils. All these factors adversely affect the life of honeybees and their population. Furthermore, 'global warming' may amplify the growth rate of pathogens like ticks, mites, viruses and fungi; many of them are called enemies of honeybees (Reddy *et al.*, 2012; Kumar and Kundal, 2016). Changes in environmental conditions and climatic scenario cause deep effects in the spread of pathogens and parasites.

Global warming also results in prolonged and comparatively hotter summer seasons. The dearth period for honeybees is extended, leading to a decrease in the bee population. Therefore, extra attention is required on the beekeeper's part to maintain colony health and strength. Beekeepers worldwide have started to notice a significant loss in the number of honeybee hives due to colony collapse disorders, with known causes such as pathogen and pests and fluctuating environmental conditions due to increasing pollution, the greenhouse effect and global warming (https://en.wikipedia. org/wiki/Colony_collapse_disorder). The results of global warming are affecting not only the wild/natural bee colonies but also dangerous for commercial beekeeping. The influence may either be directly due to heavy mortality due to extreme weather conditions or indirect due to the absence of bee flora. In this chapter, efforts have been made to review the available literature on these critical aspects and the correlation between honeybees and environmental degradation, pollination and crop productivity.

Honeybee as Potential Crop Pollinators

Honeybees are the important pollinators of flowering plants, including fruits, vegetables and other crop plants, due to which they are also known as ***angels of agriculture***. They indicate a precocious warning of altering environmental conditions and can assess various anthropogenic changes throughout long periods in an attractive way. Different factors affect the population of honeybees, including loss of habitat, fragmentation and degradation, high temperature, scarcity of flora, parasite and disease, exploitation, loss of cascades, global warming and climate changes (Le Conte and Navajas, 2008).

The conservation of bees is of immense importance to plant biodiversity as well as humankind. Successful agriculture relies upon the sowing of seeds, soil quality, temperature, moisture, pests and an abundance of pollinators. Some crops, such as mango, apples, litchi and coffee, are mass bloomed over a very short period, thus requiring a lot of pollinators for better flowering. These small pollinators are also essential for supporting wild floral biodiversity. Various types of bees, viz., 'honeybees', 'bumblebees' and 'solitary bees', are well known and constitute an economically most significant group of pollinators throughout the world; about 35% of food crop production in the world relies on these pollinators (Velthuis and Van Doorn, 2006; Klein *et al.*, 2007). It has been reported that honeybees pollinate almost 73%, flies 19%, bats 6.5%, beetles 5%, wasps 5%, birds 5% and butterflies and moths pollinate 4% of the global cultivated crops (Abrol, 2009). The dwindling of bee pollinators is a critical issue for agriculturists worldwide (Pannure, 2016). The decline in pollination due to the tremendous decrease in honeybee population has already been reported in North America and Europe. Brown and Paxton (2009) have reported that loss of habitat, pests and diseases, chemical pesticides, heavy metals and climate change can potentially impact the bee population. Many strategies were suggested, i.e., minimising the loss of habitat and its reconstruction, the prudent use of pesticides, organic farming, DNA barcoding, reducing the influence of invasive plants, animals and parasites and pathogens (Figure 9.1).

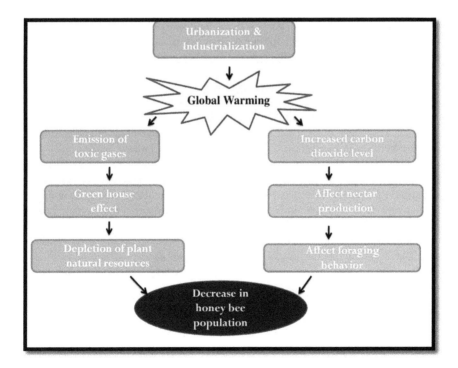

FIGURE 9.1 The impact of global warming on honeybees.

A study carried out in Britain and the Netherlands revealed that local bee diversity has a significant effect on plant production. The study's findings strongly suggest a fundamental connection between the extinction of plants and pollinator species. Change in climatic conditions may pose a serious threat to pollination services (Memmott *et al.*, 2007; Hegland *et al.*, 2009; Schweiger *et al.*, 2010). It has been reported that global temperature may increase up to 1.1°C during the 21st century, and this temperature rise shall be more in the regions of higher latitudes (IPCC, 2007). This increase in temperature has different biological impacts, and the level of impact depends upon the physiological sensitivity of organisms. The pollinators are key elements of biodiversity globally and provide essential ecosystem services to crops and wild plants. Many researchers have observed and reported that a decrease in wild and domestic pollinators' population is directly proportional to the plants which rely upon them. Decreased pollination due to the loss of pollinators has severe ecological and economic impacts that may adversely affect the diversity of wild plants, stability of the ecosystem, food production and security and human welfare. Environmental pollution directly influenced the physiology and behaviour of honeybees. Climate change can impact honeybees and flowering plants at different levels. It has been revealed that the phenology of flowering plants may get modified due to climate change. The plants either do not bear flower, possess fewer flowers or flowering time may vary due to climate change. Pollinators may not be available or may not get attracted towards such plants, which adversely affects bee-plant interaction (Thuiller *et al.*, 2005, Stokstad, 2007).

Honeybees may not get pollen and nectar at the right time, enforcing beekeepers to change their management practices. The agents that significantly contribute to pollination are flies, bees, bats, wasps, beetles, birds, butterflies and moths. However, except for honeybees, there is no single pollinator that can be managed for pollination purposes. Domesticated species of honeybees can predominantly be managed as pollinators to increase agricultural productivity. Compared to other bee species like stingless bees, honeybees can enhance crop productivity to 96% in animal-reliant pollinated crops (Klein *et al.*, 2007). Honeybees play an important role in plant pollination, especially those plants which require a true pollinator for fertilisation, i.e., bearing separate male and female flowers like an apple flower. In India, owners of apple orchards also rent bee colonies for effective pollination (Thakur and Mattu, 2015).

Apis mellifera has been considered the most important pollinator among honeybee species due to its potential to adapt to diverse climates. This species is found almost in every part of the world and most resistant to changing environmental conditions like temperature. Le Conte and Navajas (2008), in a study, reported that climate change has potential impacts on the behaviour and distribution of bees. It also affects bee communication and interaction, and orientation. Therefore, conservation measures should be used to protect natural biodiversity (Figure 9.2).

Though honeybees are the most important pollinators, their contribution is not much well estimated and documented. The reduction of 59% in the domestic stock of honeybees in the United States between 1947 and 2005 (Van Engelsdorp *et al.*, 2008; Potts *et al.*, 2010) and a 25% reduction in Europe between the year 1985 and 2005 (Potts *et al.*, 2010) have been reported. Although the number of beehives increased approximately 45%, decline in the bee population in these areas has been observed. It has a negative impact on agriculture, and the necessity of pollinators is also increasing at a very fast speed (Aizen and Harder, 2009). Urbanisation leads to the loss

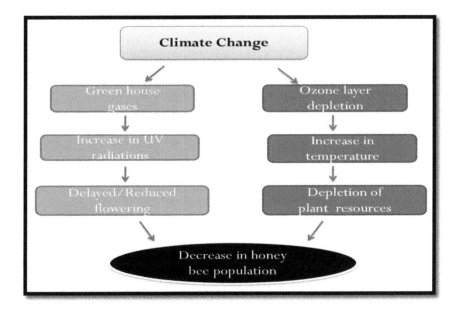

FIGURE 9.2 The impact of climate change on plants and honeybees.

of floral and nesting resources, which involves tree cutting, construction activities etc. Anthropogenic activities, habitat loss and agricultural practices have caused wide loss to pollinators (Ricketts *et al.*, 2008; Winfree *et al.*, 2009). Habitat loss also affects wild pollinator negatively. Deforestation negatively impacts wild honeybees and related processes (Steffan-Dewenter *et al.*, 2006; Brosi *et al.*, 2008). A total of 80% of the plants in Europe need insects for pollination, verifying their ecological importance (Kwak *et al.*, 1998). The decrease in the population of pollinating species, which has increased from preceding decades, may lead to a corresponding reduction of plant species or vice versa (Biesmeijer *et al.*, 2006; National Research Council, 2007; Goulson *et al.* 2008). Urbanisation, industrialisation and unsustainable agriculture negatively impact bee habitat and their abundance (Cane *et al.*, 2006; Carre *et al.*, 2009).

Increasing industrialisation, pollution and global warming may have severe impacts upon pollinators. It is expected that the activity pattern of pollinators, especially honeybees, may get changed due to an increase in temperature. Due to this, their foraging efficiency may get affected. The internal body temperature of honeybees may rise due to an increase in temperature, which could lead to overheating, followed by the death of bees. A study conducted in Bangalore (India) revealed that bee foraging activity varied significantly concerning increased temperature. Climate change may affect the biological cycle of honeybee development. Any change or shifting of honeybee from one place to another may have quantifiable consequences. Honeybees have different foraging behaviour at low and high-temperature areas. In the spring season, the queen starts egg-laying, colony develops, population increases and lots of nectar is collected and converted into honey. However, in colder areas, honeybees live tightly and use their stored food for survival, usually do not collect and produce honey in this period.

If cold weather persists for a longer period, food stores may get depleted, causing bee colonies to die of starvation (Louveaux *et al.*, 1996).

Therefore, the temperature sensitivity of pollinators should be studied to save them from heating up. Hence, proactive risk evaluation may assist us to plan against losses of pollination services due to climate change (Kjohl *et al.*, 2011). Insects are valuable pollinators and are limited in nature (Delaplane and Mayer, 2000). Farmers can manage only 11 out of 20,000 to 30,000 species of bees worldwide (Parker *et al.*, 1987), with the 'European honeybee (*Apis mellifera*)' being the most managed pollinator among all. Climate changes may lead to the growth and development of various microbes and parasites, which adversely impacts the honeybee population. Global warming negatively influences plant-pollinator interactions, phenology, abundance and distribution of both plants and pollinators. Climate change and global warming greatly affect the beginning of the flowering season and the first visit of pollinators (Hegland *et al.*, 2009).

Along with climate change, biotic stress also causes a decline in the bee population and leads researchers to look for alternative pollinators. Various bees and flies which might be the alternate of honeybees as pollinator are 'alfalfa leaf-cutter bee (*Megachile rotundata*)' and 'alkali bee (*Nomia melanderi*)' (Cane, 2002), 'mason bees (*Osmia* spp.)' (Bosch and Kemp, 2002; Maccagnani *et al.*, 2003) and 'bumblebees (*Bombus* spp.)' (Velthuis and Van Doorn, 2006). Stingless bees also prove to be significant pollinators of tropical plants, visiting more than 90 crop species (Heard, 1999). These small bees do not produce honey but visit a wide variety of crop species, making them significant pollinators (Figure 9.3).

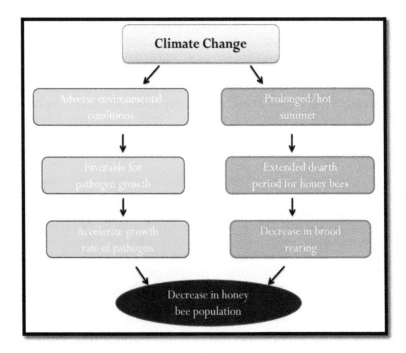

FIGURE 9.3   The impact of climate change in terms of growth of pathogen and dearth periods.

Honeybees as Indicators of Environmental Pollution

Since 1970, the bees have been employed to examine environmental pollution, especially radionuclides and heavy metals (Cavalchi and Fornaciari, 1983; Crane, 1984; Accorti *et al.*, 1986; Celli *et al.*, 1987; Stein and Umland, 1987; Celli *et al.*, 1988b) and pesticides in rural areas (Atkins *et al.*, 1981; Mayer and Lunden, 1986; Mayer *et al.*, 1987; Celli *et al.*, 1988c; Celli *et al.*, 1991; Porrini *et al.*, 1996; Wallowork-Barber *et al.*, 1982; Gattavecchia *et al.*, 1987; Tonelli *et al.*, 1990; Celli and Maccagnani, 2003; Sadeghi *et al.*, 2012; Spivak *et al.*, 2017).

Honeybees have been employed as indicators of heavy metals in urban areas since the 1960s. Pesticides and radionuclides may also be detected in the environment by analysing bee products (Celli and Maccagnani, 2003). Honeybees can also perceive the ill effects of polluting the environment and anthropogenic activities (Sadeghi *et al.*, 2012). Since the beginning of the 21st century, heavy mortality of pollinators, especially *A. mellifera*, has been reported. Researchers worldwide have mutually agreed that honeybee mortality is associated with many factors, viz., pesticides, pathogens, diseases, stress and interrupted foraging activity (Oldroyd, 2007). The traces of pesticides, fungicides were also reported in the body of adult bees and various bee products (Figure 9.4) (Celli, 1994).

Metals are the natural trace elements present on the Earth. Although humankind activities have prompted their concentration throughout the last century, in some cases, it reached the toxic range for flora, fauna and humans (Adriano, 1992). Some metals, including iron, zinc, chromium, cobalt, copper, molybdenum and manganese, are components of various chemical compounds (Peplow, 1999). These can

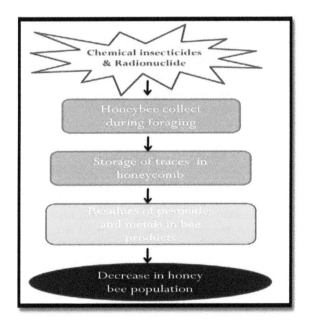

FIGURE 9.4 The impact of climate change on honeybee's w.r.t heavy metals and insecticides.

be harmful if they cross above the optimal level (Codex Alimentarius Commission, 1993). Other metals, like Ar, Pb, Cd and methyl forms of Mg, show no biological roles, and their presence causes pollution in the environment. These elements are harmful to human beings and other living organisms, even if present at very low concentration. These elements can bind with various cellular structures and inhibit their fundamental functions (Jarup, 2003). The botanical origin and composition of soil on which the plants grow is presented by the quantity of honey and various elements present in honey (González-Miret *et al.*, 2005; Lachman *et al.*, 2007). Various pollutants and their traces may get deposited on the external body surface while collecting pollen and nectar, thus brought to hive along with. These pollutants get absorbed inside the hive along with other bee products. A study conducted by Porrini *et al.* (2003) confirmed the presence of heavy metals in honey. Forty-three samples of honeybees from 16 different hives and 74 samples of honey from 29 hives were taken from industrial, urban and natural areas. The samples were examined for the presence of heavy metals. The results revealed the presence of a higher amount of chromium, lead and nickel in the samples collected from all three environments (urban: $p<0.05$; industrial: $p<0.005$; natural: $p<0.005$). In another study, different heavy metals like mercury, barium, calcium, iron, manganese, lithium, arsenic, sodium and potassium were analysed in the bee bodies obtained from Kurdistan. The results revealed that potassium was highest in the samples, followed by sodium, calcium, mercury, barium, iron, manganese, lithium and arsenic. However, there was a statistical difference among the metals analysed in these samples. Potassium was in the range of 41.857–47.871, sodium 12.653–16.183 ppm, Ca with a range of 9.077–10.058 ppm, mercury 1.12–4.786 ppm, barium 2.881–3.481 ppm, iron 1.050–1.727 ppm, manganese 0.262–0.399, lithium 0.043–0.101 and finally As in the range of 0.017–0.068 ppm. It observed that bees could collect these metals from the environment through one or another mean (Sadeghi *et al.*, 2012). Bee products like honey, wax, propolis and pollen could be examined for the presence of heavy metals, viz., Cd, Cr and Pb. The results revealed significant differences between the limits of these metals (analysed and allowable), which confirms that the honeybees are the bioindicators of environmental pollution (Conti and Botre, 2001). Metals also get introduced in the environment through natural processes like rocks weathering, volcanoes and the formation of soil (Bojakowska, 1995). Various human activities like road transportation (Kaniuczak *et al.*, 2003), agricultural activities (Klavins *et al.*, 2000) and dust coming out from urban streets (Niedzwiecki *et al.*, 2000) also affect the population of bees.

The 'colony collapse disorders (CCD)'are the principal cause that decreases the population of honeybees. Human beings are the main reason in the whole world which threatens the habitats of honeybees, and by this way, these could be used to investigate the contamination degree. The human population has threatened honeybee habitats globally, and the use of honeybees as a monitoring instrument could be well suited to investigate the degree of contamination in the environment (Seymour, 2018). When most worker bees in a particular colony disappear and leave the queen behind, with food in abundance, and some nurse bees take care of the queen and immature bees, then there is an occurrence of CCD. This disorder creates a serious threat to bees in long run. It has been already reported from different cases of CCD that honeybee colonies have substantially decreased over the last five years. In the winter season, the hive and the rate of survival of hive decreases, which is the overall indicator of bee health. In the United States, from an average of about 28.7% during 2006–2007, the population of honeybees has dropped to 23.1% for 2014–2015 (https://en.wikipedia.org/wiki/Colony_collapse_disorder). Previous studies reported that loss due to the CCD is comparatively higher during the winter season (Genersch *et al.*, 2010). A sudden loss of worker bees in a colony, parasitic mites are present in many frames in which healthy, capped brood present and shows that colonies were strong before the death of worker bees, the presence of wax moth in the beehive. A presence of a laying queen inside a beehive with other small immature bees is a prominent symptom of CCD (Caron *et al.*, 2016). Before the occurrence of CCD, the colonies of many affected beekeepers were under stress for about two months. Stresses may be poor nutrition because of the overcrowding in the apiary, crop pollination with the low value of nutrition or the dearth of pollen or nectar, pesticides exposure, supplies of contaminated water and the presence of varroa mites in high amount (Spivak *et al.*, 2017).

The activities of human beings, industrialisation, real estate and construction are leading to environmental pollution at various levels. Recent reports suggested that urban soil contamination is taking place because of the establishment of industries and urban colonies (Liu *et al.*, 2014.). It has been reported in many urban and territorial surveys that honeybees are used to observe environmental pollution caused due to heavy metals since 1970 (Cavalchi and Fornaciari, 1983; Crane, 1984), pesticides in rural areas (Gorell *et al.*, 1998; Mayer and Lunden, 1986; Mayer *et al.*, 1987; Porrini *et al.*, 1996) and radionuclides (Wallowork-Barber *et al.*, 1982; Gattavecchia *et al.*, 1987). Various types of harmful elements are present in the environment, which get transferred in plants and animals through the food chain or other means. Toxicological conditions of the natural environment affect other animals and particularly honeybees. Honeybees can sense changes in the environment, monitor the soil quality, water, plant and air pollution in areas of several square kilometres. Therefore, these small creatures act as good biological indicator (Celli and Maccagnani, 2003).

The use of pesticides at a large scale to kill the insect pests to protect the crop may have contributed to the pollinator's loss. The traces of pesticides could be detected through solid-phase extraction (SPE) and 'liquid-liquid extraction' (Rissato *et al.*, 2007; Bargańska *et al.*, 2011). These are the commonly used extraction and purification techniques to determine the residues of pesticides in honey. Honeybees can fly up to fly up to 4–5 km from the hive, but their maximum flights for up to 2 km are due to energy consumption. Based on the area foraged by bees, it has been taken into account that to produce 1 kg of honey, and there is a requirement of 100,000 foraging flights. Various bee products such as propolis, wax and pollen can be used to examine the status of environmental pollution (Celli and Maccagnani, 2003; Kalbande *et al.*, 2008). The heavy metals and radionuclides released from various sources are present throughout the biosphere and, thus, enter the food chain by accumulating in plants or uptake by the animals. The heavy metals or radionuclides pollute the surroundings, and the pollutants present in these contaminants are finally incorporated into the honeybees and their products. It may cause a deficiency in honeybees in various parts of the world (Oldroyd, 2007; Stokstad, 2007; Meixner, 2010). Various CCD factors such as biotic ones like pathogens, parasites, habitat destruction and loss, abiotic factors like climate change and contaminants affect the abundance of pollinators in the environment (Decourtye and Devillers, 2010; Carreck and Neumann, 2010; Kluser *et al.*, 2012). There is a significant increase in the use of various neonicotinoid insecticides since the introduction of imidacloprid in the early 1990s. For a significant control of crop pests in agriculture, these are widely used. These have various adverse effects on different bee species like honeybees and bumblebees. These cause various sublethal effects on reproduction, foraging behaviour, memory abilities and overwintering success (Blacquiere *et al.*, 2012). Porrini *et al.* (2002) reported that honey with radioactive metals could also act as an indicator of environmental pollution.

The Impact of Electromagnetic Radiations (EMR) on Pollinators

CCDs are the reason for the extinction of honeybees, and the other reason is the radiations that are coming from mobile phones and towers. There is a negative effect on wild bees after exposure to microwave radiations and radio waves coming from different signals. Honeybees have special organs for detecting and sensing electromagnetic forces. It has been studied in an experiment that when mobile phones are placed inside the honeybee hive with an average of 25 nT of electromagnetic waves, they disturb the life cycle of honeybees and affect their reproduction and production of honey (Nashaat *et al.*, 2013). Another study was done on electromagnetic waves coming from telecommunication towers on wild pollinators in their natural habitats. It has been reported that it has a different effect on the abundance of various pollinators (Lázaro *et al.*, 2015). A study was conducted on the impact of 'electronic medical record (EMR)' from telecommunication antennas on wild pollinators such as wild bees, bee flies, beetles, butterflies and wasps (Lázaro *et al.*, 2016). The effect of electronic medical record at various (50, 100, 200 and 400 m) distances was measured in Greece about insect abundance and richness. Electronic medical record affects all the insect pollinators except butterflies. The population of pollinators got decreased, whereas an increase was observed in the underground nesting of wild bees and bee flies. The effect of electromagnetic radiations of cell phone towers on the development of Asiatic honey, '*Apis cerana F.*', was studied in India. The colonies of *Apis cerana* were kept at 100, 200, 300, 500 and 1000 m distances from cell phone towers. The results revealed that different colony attributes, like the brood area and honey and pollen storage, were most affected in the colonies placed nearest to the towers (Taye *et al.*, 2018). The effect of 'artificial light at night (ALAN)' and 'anthropogenic radiofrequency electromagnetic radiation (AREMR)' wireless technologies (4G, 5G) is a growing threat to pollinators (Vanbergen *et al.*, 2019).

Conclusion

Urbanisation, industrialisation, anthropogenic activities and excessive use of mobiles have serious effects on flora, fauna and human health. Insects are potential pollinators, and honeybees alone are responsible for more than 80% of the pollination worldwide. The declining population of honeybees is a serious issue. The colony collapse disorders like pesticide poisoning, radionuclides, heavy metals, habitat destruction, forest fires and EMR from cell phone towers negatively affect the bee population. These events might also have additional ecological and economic impacts on the maintenance of pollinator abundance, plant diversity, crop production and human welfare. Therefore, there is an urgent need to recognise both minor and major pollinators, viz., honeybees, wasps, butterflies, hoverflies, beetles, bats, moths, birds, animals, and frame policies to conserve these pollinators to save our mother nature.

REFERENCES

Abrol DP (2009) Plant-pollinator interactions in the context of climate change – an endangered mutualism. J Palynol 45:1–25.

Vanbergen AJ, Potts SG, Vian A, Malkemper EP, Young J, Tscheulin T (2019) Risk to pollinators from anthropogenic electro-magnetic radiation (EMR): evidence and knowledge gaps. Sci Total Environ 695:133833. https://doi.org/10.1016/j.scitotenv.2019.133833

Adriano DG (1992) Biogeochemistry of trace metals. Lewis Publishers, London, Tokyo.

Aizen MA, Harder LD (2009) The global stock of domesticated honey bees is growing slower than agricultural demand for pollination. Curr Biol 19:1–4.

Al-Akhras MA, Albiss BA, Alqudah MS, Odeh TS (2015) Environmental pollution of cell-phone towers: Detection and analysis using geographic information system. Jordan J Earth Environ Sci 7:77–85.

Atkins EL, Kellum D, Atkins KW (1981) Reducing pesticide hazards to honey bees: mortality prediction techniques and integrated management strategies. Div Agric Sci Leaf 2883:1–23.

Barganska Z, Slebioda M, Namiesnik J (2013) Pesticide residues levels in honey from apiaries located of Northern Poland. Food Control 31:196–201.

Biesmeijer JC, et al. (2006) Parallel declines in pollinators and insect-pollinated plants in Britain and the Netherlands. Science 313:351–354.

Blacquiere T, Smagghe G, Van Gestel CA, Mommaerts V (2012) Neonicotinoids in bees: a review on concentrations, side-effects and risk assessment. Ecotoxicology 21:973–992.

Bojakowska I (1995) Wpływ odprowadzania ścieków na akumulację metali ciężkich w osadach wybranych rzek Polski.

Bosch J, Kemp WP (2002) Developing and establishing bee species as crop pollinators: the example of Osmia spp. (Hymenoptera: Megachilidae) and fruit trees. Bull Entomol Res 92:3–16.

Brosi BJ et al. (2008) The effects of forest fragmentation on bee communities in tropical countryside. J Appl Ecol 45:773–783.

Brown MJ, Paxton RJ (2009) The conservation of bees: a global perspective. Apidologie 40:410–416.

Cane JH (2002) Pollinating bees (Hymenoptera: Apiformes) of US alfalfa compared for rates of pod and seed set. J Econ Entomol 95:22–27.

Cane JH et al. (2006) Complex responses within a desert bee guild (Hymenoptera: Apiformes) to urban habitat fragmentation. Ecol Appl 16:632–644.

Caron DM, Togerson K, Breece C, Sagili RR (2016) The small hive beetle: a potential pest in honey bee colonies in Oregon. Oregon State University, Extension Service, Corvallis, OR.

Carre G et al. (2009) Landscape context and habitat type as drivers of bee diversity in European annual crops. Agric Ecosyst Environ 133:40–47.

Carreck N, Neumann P (2010) Honey bee colony losses. J Apic Res 49:1–6.

Cavalchi B, Fornaciari S (1983). Api, miele, polline e propoli come possibili indicatori di un inquinamento da piombo e fluoro-Una esperienza di monitoraggio biologico nel comprensorio ceramico di Sassuolo-Scandiano. In: Atti del seminario di studi "i biologi e l'ambiente" Nuove esperienze per la sorveglianza ecologica. (Manzini, P. and Spaggiari, R., Eds.) Reggio Emilia, Italy, pp 17–18; 275–300.

Celli G (1994) L'ape come indicatore biologico dei pesticidi. In: Atti del convegno: "L'ape come insetto test dell'inquinamento agricolo" P.F. "Lotta biologica e integrata per la diffuse delle colture agrarie e delle piante forestali" March 28, 1992, Florence, Italy. (D'Ambrosio, M.T. and Accorti, M., Eds.) Ministero Agricoltura e Foreste, Rome, Italy, pp 15–20.

Celli G, Maccagnani B (2003) Honey bees as bioindicators of environmental pollution. B Insect, 56:137–139.

Celli G, Porrini C, Baldi M, Ghigli E (1991). Pesticides in Ferrara Province: two years' monitoring with honey bees (1987–1988). Ethol Ecol Evol 1:111–115.

Celli G, Porrini C, Frediani D, Pinzauti M (1987). Api e piombo in città (nota preventiva). In: Atti del convegno: "Qualità dell'aria indicatori biologici api e piante". (Bufalari V., Ed.) Firenze, Italy, pp 11–45. http://www.bulletinofinsectology.org/pdfarticles/vol56-2003-137-139celli.pdf

Celli G, Porrini C, Siligardi G, Mazzali P (1988b) Le calibrage de l'instrument abeille par rapport au plomb. In: Proceedings XVIII International Congress of Entomology, Vancouver, p 467.

Celli G, Porrini C, Tiraferri S (1988c) Il problema degli apicidi in rapporto ai principi attivi responsabili (1983–1986). In: Atti Giornate Fitopatologiche, Insectology, Lecce, Italy, 2, pp 257–268. ISSN 1721-8861.

Codex Alimentarius Commission (1993) Codex guidelines for the establishment of a regulatory programme for control of veterinary drug residues in foods. CAC/GL, 16:1–46.

Conti ME, Botre F (2001) Honeybees and their products as potential bioindicators of heavy metals contamination. Environ Monit Assess 69:267–282.

Crane E (1984) Bees, honey and pollen as indicators of metals in the environment. Bee World 55:47–49.

Decourtye A, Devillers J (2010) Ecotoxicity of neonicotinoid insecticides to bees. In: Insect nicotinic acetylcholine receptors. Springer, New York, NY, pp 85–95.

Delaplane KS, Mayer DF (2000) Crop pollination by bees. CABI, New York, NY.

Gattavecchia E, Tonelli D, Bosco P (1987) Evaluation of enzymatic kinetic parameters by thin-layer chromatography with radiometric detection. Anal Chim Acta 196:259–265.

Genersch E, Von Der Ohe W, Kaatz H, Schroeder A, Otten C, Buchler R, Berg S, Ritter W, Muhlen W, Gisder S, Meixner M (2010) The German bee monitoring project: a long term study to understand periodically high winter losses of honey bee colonies. Apidologie 41:332–352.

González-Miret P, Terrab ML, Hernanz A, Fernandez-Recamales D, Heredia FJ (2005) Multivariate correlation between color and mineral composition of honeys and by theirs botanical origin. J Agric Food Chem 53:2574–2580.

Gorell JM, Johnson CC, Rybicki BA, Peterson EL, Richardson RJ (1998) The risk of Parkinson's disease with exposure to pesticides, farming, well water, and rural living. Neurology 50:1346–1350.

Goulson D, Lye GC, Darvill B (2008) Decline and conservation of bumble bees. Annu Rev Entomol 53:191–208.

Heard TA (1999) The role of stingless bees in crop pollination. Annu Rev Entomol 44:183–206.

Hegland SJ, Nielsen A, Lázaro A, Bjerknes AL, Totland Ø (2009) How does climate warming affect plant pollinator interactions? Ecol Lett 12:184–195.

IPCC (2007) Climate change 2007: synthesis report - contribution of Working Groups 1, 2 and 3 to the Fourth Assessment Report of the Intergovernmental Panel on Climate Change. In: Change IPoC, Geneva.

Jarup L (2003) Hazards of heavy metal contamination. Br Med Bull 68:167–182.

Kalbande D, Dhadse S, Chaudhari P, Wate S (2008) Biomonitoring of heavy metals by pollen in urban environment. J Environ Monit Assess 138:233–238.

Kaniuczak J, Trąba G, Godzisz J (2003) Zawartość ołowiu i kadmu w glebach i roślinach przy wybranych szlakach komunikacyjnych regionu zamojskiego. Zesz Probl Post Nauk Rol 493:133–138.

Kjohl M, Nielson A, Stenseth NC (2011) Potential effects of climate change on crop pollination. FAO, Rome.

Klavins M, Briede A, Rodinov V, Kokorite I, Parele E, Klavina I (2000) Heavy metals in rivers of Latvia. Sci Total Environ 262:175–183.

Klein AM, Vaissiere BE, Cane JH, Steffan-Dewenter I, Cunningham SA, Kremen C, Tscharntke T (2007) Importance of pollinators in changing landscapes for world crops. Proc R Soc London B Biol Sci 274:303–313.

Kluser L, Kleiber P, Holzer-Popp T, Grassian VH (2012) Desert dust observation from space—application of measured mineral component infrared extinction spectra. Atmos Environ 54:419–427.

Kujawski MW, Namiesnik J (2008) Challenges in preparing honey samples for chromatographic determination of contaminants and trace residues. TRAC-Trend Anal Chem 27:785–793.

Kumar R, Kundal N (2016) Beekeeping status in Kangra district of Himachal Pradesh. J Entomol Zool Stud 4:620–622.

Kwak MM, Velterop O, Van Andel J (1998) Pollen and gene flow in fragmented habitats. Appl Veg Sci 1:37–54.

Lachman J, Kolihova D, Miholova D, Kosata J, Titera D, Kult K (2007) Analysis of minority honey components: possible use for the evaluation of honey quality. Food Chem 101:973–979.

Lázaro A, Chroni T, Tscheulin J, Devalez C, Matsoukas, Petanidou T (2016) Electromagnetic radiation of mobile telecommunication antennas affects the abundance and composition of wild pollinators. J Insect Conserv 20:315–324.

Lázaro A, Tscheulin Th, Chroni A, Devalez J, Matsoukas Ch, and Petanidou Th (2015). Effect of mobile telecommunication antennas on the abundance of wild pollinators. Proceedings of the 14th International Conference on Environmental Science and Technology. Rhodes, Greece, 3–5 September 2015 CEST2015_01386.

Le Conte Y, Navajas M (2008) Climate change: impact on honey bee populations and diseases. Rev Sci Tech 27:485–497.

Liu Y, Su C, Zhang H, Li X, Pei J (2014) Interaction of soil heavy metal pollution with industrialisation and the landscape pattern in Taiyuan city, China. PLoS One 9(9): e105798, 1–14. https://doi.org/10.1371/journal.pone.0105798

Louveaux J, Albisetti M, Delangue M, Theurkauff J (1996). Les modalités de l'adaptation des abeilles (Apis mellifica L.) au milieu naturel. Ann. Abeille 9:323–350.

Maccagnani B, Ladurner E, Santi F, Burgio G (2003) Osmia cornuta (Hymenoptera, Megachilidae) as a pollinator of pear (Pyrus communis): fruit- and seed-set. Apidologie 34:207–216.

Mayer DF, Johansen CA, Lunden JD, Rathbone L (1987) Bee hazard of insecticides combined with chemical stickers. Am Bee J 127:493–495.

Mayer DF, Lunden JD (1986) Toxicity of fungicides and an acaricide to honey bees (Hymenoptera: Apidae) and their effects on bee foraging behavior and pollen viability on blooming apples and pears. Environ Entomol 15:1047–1049.

Meixner MD (2010) A historical review of managed honey bee populations in Europe and the United States and the factors that may affect them. J Invert Pathol 103:S80–S95.

Memmott J, Craze PG, Waser NM, Price MV (2007) Global warming and the disruption of plant-pollinator interactions. Ecol Lett 10:710–717.

Nashaat el halabi, Gaby Abou Haider and Roger Achkar (2013) The Effect of Cell Phone Radiations on the Life Cycle of Honeybees. In IEEE EUROCON 2013. DOI:10.1109/EUROCON.2013.6625032.

National Research Council (2007) Status of pollinators in North America. The National Academies Press, Washington, DC. https://doi.org/10.17226/11761.

Niedzwiecki E, Protasowicki M, Kujawa D, Niedzwiecka D (2000) Zawartość kadmu i ołowiu w pyle opadowym w obrębie aglomeracji szczecińskiej [w: Kadm w środowisku-problemy ekologiczne i metodyczne]. Zeszyty Naukowe Komitetu "Człowiek i środowisko" PAN 26:201–208.

Oldroyd BP (2007) What's killing American honey bees? PLoS Biol 5:168.

Pannure A (2016) Bee pollinators decline: perspectives from India. Int Res J Natl Appl Sci 3:2349–4077.

Parker FD, Batra SWT, Tepedino VJ (1987) New pollinators for our crops. Agricult Zool Rev 2: 279–304.

Peplow D (1999) Environmental impacts of mining in Eastern Washington. University of Washington Water Center. http://hdl.handle.net/1773/17077

Porrini C, Colombo V, Celli G (1996) The honey bee (*Apis mellifera* L.) as pesticide bioindicator. Evaluation of the degree of pollution by means of environmental hazard indexes. In: Proceedings XX International Congress of Entomology, Springer Science & Business Media, Firenze, Italy, pp 444.

Porrini C, Ghini S, Girotti S, Sabatini AG, Gattavecchia E, Celli G (2002) Use of honey bees as bioindicators of environmental pollution in Italy. In: Honey Bees: Estimating the Environmental Impact of Chemicals. Taylor & Francis, London and New York, pp 186–247.

Porrini C, Sabatini AG, Girotti S, Ghini, Medrzycki P, Grillenzoni F, Bortolotti L, Gattavecchia E, Celli G (2003) Honey bees and bee products as monitors of the environmental contamination. Apicata 38:63–70.

Potts SG, Jacobus CB, Kremen C, Neumann P, Schweiger O, William EK (2010) Global pollinator declines: trends, impacts and drivers. Trends Ecol Evol 25:345–353.

Reddy PVR, Verghese A, Rajan VV (2012) Potential impact of climate change on honeybees (*Apis* spp.) and their pollination services in horticultural ecosystems. Pest Mange Hortic Eco 18:121–127.

Ricketts TH et al. (2008) Landscape effects on crop pollination services: are there general patterns? Ecol Lett 11:499–515.

Rissato SR, Galhiane MS, de Almeida MV, Gerenutti M, Apon BM (2007) Multiresidue determination of pesticides in honey samples by gas chromatography–mass spectrometry and application in environmental contamination. Food Chem 101:1719–1726.

Sadeghi A, Mozafari AA, Bahmani R, Shokri K (2012) Use of honeybees as bio-indicators of environmental pollution in the Kurdistan province of Iran. J Apic Sci 56:83–88.

Schweiger O, et al. (2010) Multiple stressors on biotic interactions: How climate change and alien species interact to affect pollination. Biol Rev 85: 777–795.

Seymour N (2018) Bad environmentalism: irony and irreverence in the ecological age. University of Minnesota Press. https://blogs.lse.ac.uk/lsereviewofbooks/2018/10/19/book-review-bad-environmentalism-irony-and-irreverence-in-the-ecological-age-by-nicole-seymour/

Spivak M, Browning Z, Goblirsch M, Lee K, Otto C, Smart M, Wu-Smart J (2017) Why does bee health matter? The science surrounding honey bee health concerns and what we can do about it. https://www.cast-science.org/wp

Steen JJN, Van der, de Kraker J, Grotenhuis T (2012) Spatial and temporal variation pszczół miodnych of metal concentrations in adult honeybees (*Apis mellifera* L.). Environ Monit Assess 184:4119–4126.

Steffan-Dewenter I et al. (2006) Bee diversity and plant-pollinator interactions in fragmented landscapes. In Specialization and generalization in plant-pollinator interactions. University of Chicago Press, pp 387–410,

Stein K, Umland F (1987) Mobile and immobile sampling by means of bees and birches. Fresenius' Z Anal Chem 327:132–141.

Stokstad E (2007) The case of the empty hives. Science 316:970–972.

Taye RR, Deka MK, Borkataki S, Panda S, Gogoi J (2018) Effect of electromagnetic radiation of cell phone tower on development of Asiatic honey bee, *Apis cerana* F. (Hymenoptera: Apidae). Int J Curr Microbiol App Sci 7(8): 4334–4339. doi. org/10.20546/ijcmas.2018.708.454.

Thakur B, Mattu VK (2015) Effect of honeybee pollination on quantity and quality of apple crop in Kullu Hills of Himachal Pradesh, India. Int J Sci Res 4(4): 2015.

Thuiller W, Lavorel S, Araujo MB, Sykes MT, Prentice IC (2005) Climate change threats to plant diversity in Europe. Proc Natl Acad Sci USA 102:8245–8250.

Tonelli D, Gattavecchia E, Ghini S, Porrini C, Celli G, Mercuri A (1990) Honey bees and their products as indicators of environmental radioactive pollution. J Radioanal Nucl Chem 141:427–436.

Van Engelsdorp D et al. (2008) A survey of honey bee colony losses in the U.S., Fall 2007 to Spring 2008.

Velthuis HH, Van Doorn A (2006) A century of advances in bumblebee domestication and the economic and environmental aspects of its commercialization for pollination. Apidologie 37:421–451.

Wallowork-Barber AK, Ferenbaugh RW, Gladney ES (1982) The use of honey bees as monitors of environmental pollution. Am Bee J 122:770–772.

Winfree R et al. (2009) A meta-analysis of bees' responses to anthropogenic disturbance. Ecology 90:2068–2076.

10

Biochemical and Molecular Mechanisms for Medicinal Properties of Honey

Subramani Srinivasan, Raju Murali, Veerasamy Vinothkumar,
Ambothi Kanagalakshimi, Natarajan Ashokkumar and
Palanisamy Selvaraj
Annamalai University, Annamalainagar, India

Subramani Srinivasan, Raju Murali and Ambothi Kanagalakshimi
Government Arts College for Women, Krishnagiri, India

Vinayagam Ramachandran
Yeungnam University, Gyeongsan, Korea

Devarajan Raajasubramanian
*Annamalai University, Annamalainagar, India and Thiru A.
Govindasamy Government Arts College, Tindivanam, India*

CONTENTS

Introduction

Honey is innate substance collected from flowers across the vegetation by foraging of bees. In the stomach of bees digestive enzymes act upon the sucrose of nectar and it is dissociated into glucose and fructose. Honeybees spit up their nectar into another bee's mouth and this process occurs frequently (1200 seconds) till nectar

DOI: 10.1201/9781003175964-10

further rehabilitates into fresh honey. Bees spittle the raw honey into the cells of honey comb and fold their wings to dehydrate it, then cover it with wax (Saralaya and Thomas, 2021). Honey has been used as a medicine and a natural cure for a variety of diseases in several ancient civilisations. Bees are the primary pollinators around the world, honey is bee glean nutritious treacle with viscid in nature; super sopping jelly compounds, composed of intricate constituents such as water(10–20%), sugars (70–80%) in the form of monosaccharides are found in nearly 75% as well as 10–15% disaccharides and little amount of other sugars, amino acids, carotenoids vitamins, enzymes, organic acids, phenolic compounds, phytochemicals, minerals and natural antioxidants; nearly 600 compounds have been isolated from honey (Przybylski and Bonnet, 2021). Notably, honey is encompassed of >200 elements, which are extremely reliant on the plant source usage to yield. Approximately 300 different types of honey originated is due to different in colour, flavour, sensory perception and medicinal response and may vary in chemical composition, influenced by botanical (flower), honeybee species, geographical origin, harvesting period (techniques and condition), season, type of processing and storage condition. The configuration, colour, aroma and flavour of honey are relatively flexible and be contingent on its floral basis, topographical areas and ecological aspects. Most of the chemical component honey from nectar of plants and other substances added to honey by bees. The uses of honey have been recorded since time immemorial besides depicted in paintings of prehistoric period indicating in diet, traditional medicine and have currently been rejuvenated as an outstanding therapeutic perspective. Since antiquity, honey was utilised for an assortment of affliction without understanding the health-promoting properties (Talebi et al., 2020).

History and Therapeutic Perspective of Honey

The perception of honey consumption as a medication is underway since the Stone Age and the initial suggested histories on Neolithic age medications evidently validate that honey was exploited as an oldest food known by Egyptians. Greeks with historical progression used honey in human feeding habits and as portion of emulsions healing injuries in the form of wounds. Further, in modern history even during the World War1, honey was used in battlefield to treat the wounds of warriors. Wound-healing properties of honey were reported in the United States and Europe in the middle of the 20th century. Today, subsequently, honey was used as a therapeutic agent and generally documented as a basis of human nourishment since classical times. Fortunately, development, application and also utilisation of traditional systems of medicine reduced modern antibiotics in recent years (Vogt et al., 2021). The application of traditional medicine to cure infection has been practiced since the origin of human beings, and honey produced by *Apis mellifera* is one of the oldest conventional medicines considered to be foremost in the treatment of several complications (Dai et al., 2020). Honey and their products (bee pollen, propolis, honeybee wax and royal jelly) have been used worldwide since prehistorical ages as a medicine because of the biological properties which show a lot of health benefits and therapeutic values. In past two decades, researchers have focused on socio-economic and health benefits of honey and treated as 'Sarva

Roga Nivarini'; honey in today's medicine because of their biologically active compounds and properties for treating various ailments shows a lot of promising health benefits (Huang et al., 2020).

In the human discovery, honey is commonly recognised as a 'holy grail' in several native residents of all over the earth for medical and nourishment applications. Therapeutic possessions of honey critically depend on its basic configuration that might be different on numerous factors like type of flower, environmental conditions, processing methods and species of bee. Enlightening the effect of honey in medications is crucial for vindicating its usage and confirming the quality of therapeutic nutrition. Bobis and his coworkers (2020) speculated that honey was an extortionate natural invention with excellent curative characters primarily applied as ointment to wound repair and to avert suppuration. Further, a study by Sinyorita et al. (2011) established that in Indian Ayurvedic medicine, honey is deliberated an extremely valued divine nutrient, compatible food, improves the efficiency of drug and concurrently alleviates side effects of the medication. Medicinal grade of honey is well documented in accordance with floral source, among these highly valued and priced jelly is Manuka honey derived from tree *Leptospermum* genus, which is indigenous of New Zealand and Australia shows antibacterial properties (Johnston et al., 2018). A recent work by Lumbers (2020) reported that substantiated honey has been used as agent for antimicrobial, antiviral and antiparasitic, due to low water activity, low pH and bioactive compounds promote blood distribution by angiogenic method among the tissues particularly in the area of wound; besides, formation tissue granulation and skin re-epithelialisation also reduce oxidative stress which is the prime mechanism behind the wound healing.

Honey is a centuries-old food that is commonly consumed for its therapeutic, cosmetic and nutritious benefits. Honey should be free of any food component, including food materials, and also used in any format anticipated for people ingestion. Honey is a virtuous source of vitality due to its high carbohydrate and sugar content, and it is often used in cooking as a natural sweetener. Phytochemicals are the utmost common source of bioactivity, and therapeutic scenarios are transmitted to honey by bees collecting pollen and floral nectar. They differ between plants in terms of the amount and variety of bioactive compounds in honey, but the most abundant are flavonoids, terpenoids and polyphenols, which are biologically effective and have antioxidant effects. Despite this, it is practically impossible to provide a comprehensive description of the phytochemical characteristics of honey and pollen in a literature review of this scope due to space constraints. The beneficial properties of biologically energetic efficacy, as well as honey usage in effective nutrition foodstuffs, are also discussed in this chapter. Numerous epidemiological and research evolutions have established that honey is extremely competence as a prospective therapeutic medication and can be used as remedy for innumerable human illnesses. Noteworthy, this chapter delivers evidence-based details on the pharmaceutical potential of honey along with its beneficial treatments and accurate mechanisms of activity. In this chapter, we critically focus in detail on the most recent scientific works from Medline-based literature searches and scientific libraries supports regarding the potential therapeutic perspective of honey for the treatment of diverse ailments and enlighten the promising mechanisms of efficacy in both *in vivo* and *in vitro*. Specified its scope, it is a beneficial implement for researchers and scientists involved in drug discovery and the therapeutics and pharmacology of honey.

Antioxidant Properties of Honey

Dietary antioxidants have attracted great attention to combat of various diseases. Furthermost in ancient civilisations, complete antiquity has used honey for constructive properties on human health and for its medicinal effects and its biological action. Evidence suggest that in human physiology prerogatives, free radicals are intricate in virtually most cellular breakdown mechanisms leading to cell lysis and a shocking upsurge in mortality from chronic non-communicable diseases, moreover due to free radicals communally accountable for almost 70% of deaths in the earth (Carrier, 2017). Notably, Khosla et al. (2020) established that most of these ailments, for instance, cancer, atherosclerosis, neurological disorders and diabetes mellitus (DM) are connected through an augment in oxidative stress. Moreover, Daneshzad et al. (2020) advocated that the prominence of antioxidants from natural treasure in free-radical injuries has greater insinuations in human diseases prevention and treatment. Fascinatingly, Al-Hatamleh et al. (2020) published that honey shows its antioxidant efficacy by thwarting diverse ailments associated with non-communicable and communicable illnesses. Honey is a nutritional material, comprising phytochemicals (rich source of polyphenols), non-enzymatic and enzymatic antioxidants and vitamins naturally. Honey mutually aimed in pharmaceutical as well as medical determination has come to be observed as a conventional medicine in current eras (Munstedt and Männle, 2020). Honey as antioxidant contributes an electron to the unpaired valence electron inside the free radical, consequently triggering the immune response and averting cell injury. The antioxidant activity primarily correlated to phytochemicals from honey through pollen and floral sources transported by the bee via its saliva (Habryka et al., 2020). Goslinski and his associates (2020) hypothesised that the notorious mechanisms of honey evidence to overcome oxidative impairment defends on (i) antioxidant enzymes, (ii) polyphenols as metal chelators and as outstanding free-radical scavengers (iii) and persuade cellular non-enzymatic and enzymatic antioxidant structures. Consequently, Hailu and Belay (2020) authenticated that honey supplementation benefits in the medication of chronic ailments generally concomitant with oxidative damage. Nevertheless, the antioxidant properties of honey have been substantiated in an organised method, there are numerous mysterious facets still to be originated (Figure 10.1).

Antimicrobial Efficacy of Honey

The surprising increase and rapid expansion of antimicrobial resistance are the most pernicious distresses for the human community, as well as for the water and food consumed by humans. The universal incidence of transmissible illnesses, exclusively bacterial contagions, composed of antibiotic resistance, exploited a critical people health problem, resulting in sustained infection and mortality. Although there have been significant developments in pharmaceutical applications, we continue to be confronted with novel microorganisms that pose a risk to healthcare systems, human health and universal economic safety. A report from the World Health Organization

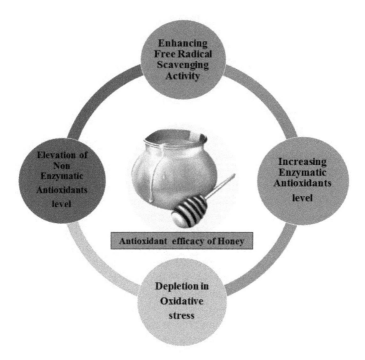

FIGURE 10.1   Graphical portrayal of antioxidant efficacy of honey.

(WHO) identifies infective ailments as the third utmost substantial origin of death worldwide and recently reported study speculates that severe acute respiratory syndrome coronavirus 2 (SARS-CoV-2)-infected persons are susceptible to progress subordinate microbial coexistence precisely sepsis and bacterial pneumonia which is vital hazard (Rawson et al., 2020). Nevertheless, the shocking augmentation in the multidrug-resistant pathogens stimulates the necessity for emerging replacement antimicrobial approaches. It is crucial to discover some alternatives that can potentially be effective in the treatment of these infectious illnesses. There is substantial evidence given by Hbibi et al. (2020) that honey has potential biological activities since antiquity for its nutritious and restorative effects, specifically for its antimicrobial characteristics. Subsequently, several pharmaceutical properties of honey predominantly diverse phytochemicals with antimicrobial efficacy, making it a powerful antimicrobial weapon to combat against microorganisms. The key ingredients mechanism of honey against human pathogens varies between Gram-positive and Gram-negative bacteria and some cellular points could be broadly specific for two types of bacteria (Ullah et al., 2021). The antimicrobial mechanism of honey has been accredited to several factors, including pH, osmolarity, phenolic components, flavonoids and particularly hydrogen peroxide (H_2O_2).

Honey's potential antimicrobial efficacy and antiviral action are probably due to H_2O_2. In addition, intake of honey (presence of H_2O_2) might provide a protective measure, might assist to sanitise the throat from infected virus elements. Blackman et al. (2020) speculated that the involvement of H_2O_2 to the antibacterial efficacy has been evaluated, they established that H_2O_2 attenuation through catalase potentially diminishes the antibacterial effectiveness of honey. The H_2O_2 in honey is formed thru the activity of the enzyme glucose oxidase, and it is synthesised into accumulated nectar from the hypopharyngeal gland of the bees and it is triggered through the oxidation of glucose into gluconic acid and this acid stimulates membrane depolarisation and chronic susceptibility to gluconic acid caused bacterial membrane devastation (Bang et al., 2003). As revealed by Masoura et al. (2020), oxidative stress, which produced hydroxyl radical, has been exposed to be the crucial mechanism of H_2O_2 in honey influencing sequence of mechanisms associated to progress origination, cell partition and cell wall synthesis, altogether vital for bacterial feasibility. However, hydrolysis of H_2O_2 can also produce oxygen (O_2), which can augment autoxidation of other phytochemical compounds in honey, to stimulate pro-oxidant agents leading to synthesis additional H_2O_2. Fascinatingly, the breakdown of H_2O_2 is catalysed by ions especially Fe^{2+} or Cu^+, existing in honey, synthesis hydroxyl radicals and the dissemination of radicals via bacterial cell wall and destroys the cell wall veracity, which are more responsible for antimicrobial efficacy (Bouzo et al., 2020). Despite this, H_2O_2 can react with benzoic acid to create peroxy acids, which are more stable and powerful antimicrobial elements than H_2O_2 and can counteract catalase.

While H_2O_2 is extensively contemplated as the prime origin of the antimicrobial efficacy of honey, there are diversities the key antibacterial characteristics of which are non-H_2O_2 reliant. Besides H_2O_2, antibacterial efficiency of honey is powerfully interrelated with its natural physicochemical features predominantly viscosity, thickness, pH and osmolarity (Yu et al., 2020). The antibacterial properties of honey are partly due to acidity, with a pH from 3.5 to 4.5 and this acidic pH exploits its antimicrobial effect. Subsequently, honey's high osmolality suppresses bacterial growth by removing moisture from the atmosphere, which will desiccate bacteria. An alternative mechanism of honey to stop the spreading of infection is capability of honey to stop cohesion of bacteria to cells. Wang et al. (2021b) described that susceptibility of *Salmonella enteritidis* to honey for 1 hour earlier to fraternisation the fresh bacteria with intestinal epithelial cells diminished the quantity of bacteria combined to the cells by 74%. Further, bacterial membrane potential plays a crucial part in numerous bacterial physiological progressions and the loss of the equipoise of ion concentration in and out of bacteria might disturb to their feasibility. According to Combarros-Fuertes et al. (2020) findings, honey augments the penetrability of the outer membrane of Gram-negative bacteria by eradicating the lipopolysaccharide layer and also modifies the membrane efficacy in both Gram-positive and Gram-negative bacteria. Natural honeys are rich in phytochemicals; recently Das and his colleagues (2020) discovered that they forecast binding affinity of naturally existing honey phytochemicals to SARS CoV-2 viral proteins which are led in the context of computational and docking studies, and an anticipated medicinal value of phytochemicals is based on their predictable binding to SARS-CoV-2 main protease. All these properties advocate that honey can play a momentous role in clinical practice in antimicrobial therapies to treat microorganisms. The mechanistic efficacy of honey and the bacterial cell retorts to it would provide a foundation for the creation of honey-originated therapeutic modules with enhanced antibacterial competence (Figure 10.2).

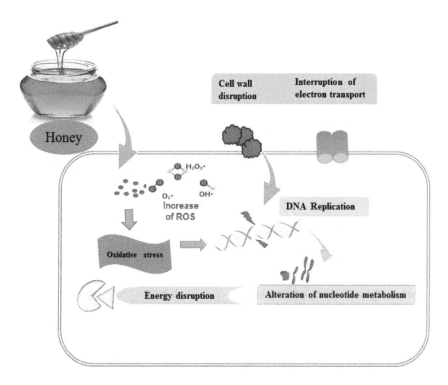

FIGURE 10.2 Antimicrobial mechanism of honey against human pathogens.

Antidiabetic Effects of Honey

DM is a disease in which blood glucose levels remain high. The incidences of DM are rising fast, and treatment is a tremendous challenge worldwide. Honey is a regular sweetener that has been used in common food and its medicine activities. The present work indicated that honey administration (2.0 g/kg body weight [BW]) resulted from the reduction synthesis of protein and fats, which are transformed into vitality (Sahlan et al., 2020). Pieces of evidence provided by clinical studies reported that honey consumption reduces the postprandial glycemic response and glucose levels in serum of type 1 and 2 diabetic patients (Shambaugh et al., 1990). A study by Sadeghi et al. (2020) reported that honey treatment to type 2 diabetic patients was reduced malondialdehyde, C-reactive protein, low-density lipoprotein cholesterol (LDL-C), adiponectin and substantially augmented high-density lipoprotein cholesterol (HDL-C). Many preclinical and clinical reported that the expected honey consumption statement recover glycemic normal in mice and humans affected by DM decreased the levels of glycated haemoglobin (HbA$_{1C}$) and improved BW in diabetic patients (Nazir et al., 2014).

Further study confirmed that antioxidant activity of honey helps improve the oxidative lipid metabolism in type 2 DM patients (Rahimi et al., 2005). The research work published in 2019 by El-Haskoury et al. examined the ethyl acetate honey extract decreased significant hepatic and lipid profile in diabetic rats. Batumalaie et al. (2013) studied the Gelam honey's beneficial effect on diabetes' insulin signalling and inflammation mechanism. Treatment with Gelam honey improves insulin content, increased insulin receptor substrate 1 (IRS-1), phosphorylated protein kinase B (also known as p-Akt), increased pro-inflammatory cytokine such as tumour necrosis factor-alpha (TNF-α), interleukin-6 (IL-6) and interleukin-1 beta (IL-1β) in HIT-T5 cells (Batumalaie et al., 2013). The study of Batumalaie et al. (2014) confirmed the treatment of HIT-T15 cells by Gelam honey significantly reduced the reactive oxygen species (ROS), lipid peroxidation and increased levels of insulin levels. These attenuated the oxidative-induced inflammatory pathways when treatment with honey was downregulation of p-Jun-N-terminal kinase, and pro-inflammatory cytokines include IL-6, TNF-α, IL-1β and increased Akt (protein kinase B), representing the preventive effect against inflammation and insulin resistance (Safi et al., 2016) (Figure 10.3).

In research conducted by Aziz et al. (2017), the stingless bee honey orally treatment to diabetic animals prevents high levels of fasting blood glucose levels, cholesterol, triglycerides, LDL-C. However, insulin and HDL-C levels in diabetic rat's administration stingless honey increased. In this context, treatment with honey was significantly decreased glucose levels of streptozotocin/alloxan-induced diabetic rats (Fasanmade and Alabi, 2008). Erejuwa et al. (2010a) explored honey has been reduced glucose levels in serum of experimental diabetic rats. The same author has proven that honey's potent antioxidant property improves insulin secretion and reduced oxidative stress (Erejuwa et al., 2010b). Evidence provided by Folli et al. (2011) suggest that honey can affect antioxidant ability concerning its dose: high ROS production,

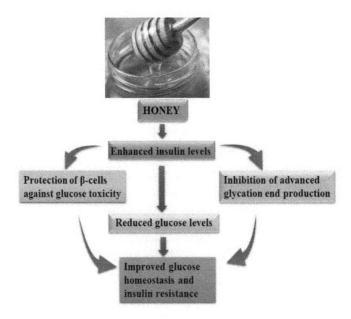

FIGURE 10.3   Effect of honey on maintenance of glucose homeostasis in diabetic state.

synthesis of glycogen and glucose uptake and impairment of insulin signalling pathway, which the honey supplement can restore.

Chemotherapeutic Potential of Honey

One of the universal pandemic is invasiveness of malignancy. Globally around 10 million people are identified with malignancy every year and over half of the affected persons ultimately passed on from the ailment; moreover, it is amongst the top five foremost origins of worldwide mortalities (Rahman et al., 2020). Arora and his research team (2020) is projected that there were 18.1 million prevalence cases of malignance worldwide with 9.6 million deaths in 2018 and in India, in the similar year, approximately 1.15 million malignancy occurrence and 7.84 lakh deaths were documented. Report of Kocarnik (2020) estimated that 17.0 million new cancer incidences in global level, of which 657,000 cases take place in low Human Development Index (HDI) countries, 2.8 million cases in countries with medium HDI, 6.4 million incidences in high HDI. WHO assesses approximately 15 million new candidates of malignancy by 2020. Noteworthy, in India, malignancy is the second utmost customary ailment answerable for the highest death and the statistics also reveal in India cancer origins in excess of 10% of the annual mortalities in India (Chauhan et al., 2020). In spite of extensive investigate effort, this invasive circumstance remains a provocation to avert and treat. Current cancer exploration suggests inventive treatment approaches that have controlled the malignancy results. Conservative approaches to manage tumour have serious hazards requiring the exploration for innovative, less toxic remedies. More recently, there has been augmented attention in the investigation for chemoprevention as well as in tumour rehabilitation agents derived from dietary molecules and nutraceuticals (Farhana et al., 2020). Nutritional ingredients as well as natural foods have been confirmed to ameliorate regular signalling pathways in malignancy growth. The 'natural cancer vaccine' – honey – can able to diminish unrelieved inflammatory processes, progress immune response and decrease the infections by strong organisms. This chapter intended to pucker the evidence on proposed mechanisms for antineoplastic efficiencies of this natural food in the published reports (Figure 10.4).

Scientific investigations advocate that putative role of cancer cells to ROS and greater relationship of lipid peroxidation by-products might be competence in malignance cell injury. The assumed part of oxidative stress in stimulating the oncogenic progression through free radicals and the defensive efficacy of antioxidants by scavenging free radical in malignancy growth are well established by Wang et al. (2021a). However, honey is sturdily documented as a free-radical hunter along with sturdy antioxidant and consequently, cumulative evidence displays that honey may be a significant chemopreventive molecule via attenuating ROS and ameliorating apoptosis (Martinotti et al., 2020). Furthermore, amplified information advocates that the antioxidant efficacy of honey potentially displays a vital result in averting the instigation and the augmentation of cancer. Research study evolved universally has evidenced that honey has better implications in tumour alleviation due to inhibition of three key phases of malignancy, initiation, propagation and metastasis. This chemopreventive efficacy has been accredited to proapoptotic mechanisms, ameliorations of oxidative

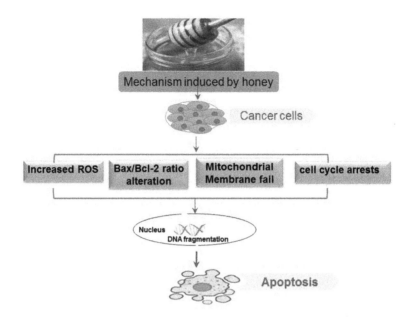

FIGURE 10.4 Mechanism of chemotherapeutic efficacy of honey.

stress, cell cycle inhibition, stimulation of mitochondrial signal cascade and angiogen-esis inhibition (Al-Koshab et al., 2020). Immunomodulatory and immune defensive perspective has higher insinuations in malignancy prevention and treatment of honey. The aforementioned statement proved by recent scientific exploration that honey originates the B-cells, macrophages and T-cells to better significances in cancer pre-vention and besides, non-sugar constituents of honey might responsible for immune potentiating and immunomodulation action (Badolato et al., 2017). Afrin and his associates (2020) exhibited that the mechanism for anti-inflammatory effect of honey implicates the obstructing of the pro-inflammatory actions of stimulatedcyclooxygen-ase-2 (COX-2) and inducible nitric oxide synthase (iNOS). Honey also documented to be intricate in the amelioration of iNOS, tyrosine kinase, ornithine decarboxylase and COX-2. Honey has the potential to obstruct eventual carcinogenesis steps but also attenuate the inflammation rate and leads to affect tumour propagation. However, Ahmed and Othman (2013) suggested that honey is imperative to hinder the cell cycle of glioma, melanoma in colon malignance cell lines in GAP 1/GAP 0 stage. As well as, this natural food augments the caspase 3 and different proapoptotic proteins, while it has been stated to down-regulate the antiapoptotic protein, B-cell leukaemia 2 (Bcl-2) and upsurges the appearance of proapoptotic protein Bcl 2-associated X protein (Bax). In light of the previously disclosed reality, honey's apoptotic prowess clearly qualifies it as a potential cancer-fighting nominee (Abubakar et al., 2012). Based on aforementioned declarations, honey is an optimistic pharmacological molecule for impeding human melanoma propagation. Probable randomised detailed experimental and clinical explorations are required to substantiate the validity of honey as candi-date treatment of malignancy.

Role of Honey in Alzheimer's Disease

Alzheimer's disease (AD) is congenital disease that causes permanent memory loss and the primary cause of mental illness worldwide. Research has found that honey administered was reduced oxidative stress levels, acetylcholinesterase (AChE) property, neuronal proliferation and improved memory (Azman et al., 2015). This result was evinced that honey (1–1.5 ml/kg BW) treat was altered neurogenesis in the hippocampus and AChE activity (Azman et al., 2016). The latest report is that honey treatment inhibits brain and cholinergic inhibitors and protected AD-like behavioural disturbances (Baranowska-Wojcik et al., 2020). Honey treatment significantly reduced neuroinflammation and neural apoptosis on kainic acid (KA)-induced excitotoxicity (Mohd Sairazi et al., 2017). Oral therapy with honey reduced neurodegeneration through its attenuated activity and increased KA-induced inflammatory markers, COX-2, allograft inflammatory factor 1 and caspase 3 efficiency (Mohd Sairazi et al., 2018). Research has also confirmed that honey flavonoid extract (HFE) significantly inhibited inflammatory proteins' release, such as TNF-α and IL-1β. HFE is a powerful inhibitor of microglial activity and therapeutic for neuroinflammation involving neurodegenerative disease (Candiracci et al., 2012). It has also confirmed the antinociceptive properties of diverse honey samples. Oral admiration of honey showed a more in their pain thresholds in mice. This analgesic property of honey is possibly refereed by opioid receptors (Figure 10.5) (Aziz et al., 2014).

According to Syarifah-Noratiqah et al. (2018), a polyphenol from honey decreased the release of neurotoxic ROS and the deposition of misfolded proteins. The results exhibited that honey supplementation for 1 day by H_2O_2 prevents cellular death

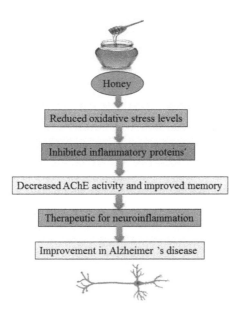

FIGURE 10.5 Therapeutic role of honey on Alzheimer's disease.

through its antioxidant properties. Othman et al. (2015) found that the development of memory and learning related brain area after honey treatment may enhance the cholinergic system, microglial activation and inhibition of neuroinflammatory. These findings evidence postmenopausal women improve in the immediate memory and reduced estrogen honey 20 mg for 16 weeks (Othman et al., 2011). Honey indicates protective activities for cognitive impairment (Rittschof et al., 2018). These findings further evidenced by Qaid et al. (2021) show that Tualang honey-treated hypoxia significantly prevents neuronal damage in sucrose-induced hypoxia. Increased memory performance, acetylcholine (Ach) concentrations and reduced AChE in the brain homogenates of stressed ovariectomised were treatment with Tualang honey in rats (Al-Rahbi et al., 2014). In recent studies, Wan Yaacob et al. (2021) reported that Tualang honey/methanol fraction significantly reduces TNF-α and COX-2 protein expression and improved spatial/memory. Meanwhile, Mohd Sairazi et al. (2018) also found a similar result that Tualang honey reduced caspase-3 and neuroinflammation after KA-induced status epilepticus and honey reduced anxiety and improved spatial memory. Yahaya et al. (2020) have performed that Tualang honey supplementation patients for 8 weeks substantially augmented tutoring output across spheres in the instant memory. Tualang honey supplementation improved memory performance and decreased AChE activity in experimental male rats (Azman et al., 2016). Oral treatment of honey shows the convulsions and sleep time were reduced, as well as improved anxiety in rats (Akanmu et al., 2011).

Therapeutic Efficiency of Honey in Wound Repair

Honey's medicinal physiognomies have long been recognised by historic beliefs, which is used by human beings since olden days. Honey is a vital sweet food from natural paragon signifying latent therapeutic characters from antique Egyptian evolution. In the Romans, Greeks and Assyrians usage of honey in wound curative and Hippocrates, an ancient Greek physician favoured honey over rugs, promoting its application for various illnesses. Due to severe side effects of modern medicinal systems, this prehistoric system of healing is nowadays under exploration. Honey has been documented as a source of nourishment and medicine from 2100 to 2000 Before Christ (BC)onwards; in spite of this, Aristotle, a Greek polymath (384–322 BC), designated this natural nutrition as 'good as a salve for wounds' (Zubai and Aziz, 2015). Hippocrates recommended this sweet food for abscesses, burns and boils, whereas Dioscorides, a surgeon of the Roman armed forces in 50 BC, used honey for wound repair. Nevertheless, the sweet food has only piqued scientific interest since the late 19th century. The antique therapy of honey for dressing infected wounds is quickly becoming re-explored in proficient medication, exclusively where injuries are infected with antibiotic-resistant bacteria. Yet, it is extremely authoritative to take serious methods towards wound precaution and curative molecules mainly its method of healing route which is an intricate mechanism and vulnerable to exterior infections. Nonetheless, the current therapeutic works have revealed a modification in the tendency with numerous research publications explored its efficiency in treating diverse types of wounds (Yilmaz and Aygin, 2020). Honey is used for pharmacological therapeutic purposes since olden days in wound treatment, it is enormously crucial to sound the therapeutically effectiveness of honey in respect to the physiology of the injured skin in order to augment the curative management.

Honey was initially portrayed in antiquity as an appreciated nourishment source, at a time when the globe had not yet revealed sugar and widely used for wound-healing purposes. The beneficial efficacy of honey in wound repair is crucially based on its antibacterial, antioxidant and anti-inflammatory features all are expansively deliberated in the last10 years (Nizet et al., 2020). Honey has all constructive principles to be used in wounds healings and is a virtuous alternate to synthetic drugs. Nevertheless, extensive scientific works revealed therapeutic efficiency of honey in wound repair, the plausible mechanisms are still mostly unclear. We showed in this chapter to review the role of honey as an effective intervention for the treatment of wound. As necrotic tissues can lead to progression of infecting microorganisms and persuade wound to develop sustained, it is imperious to eliminate any dead tissue (Martinotti et al., 2019). Besides, honey's high viscosity supports as a physical barrier that prevents the entry of microorganisms and retains the wound humid to strong curative, further to averting scab development and damaged external tissue. Noteworthy, one of the mechanisms of honey hastens restorative actions through initiating antioxidant activities, decreasing inflammation, autolytic debridement and immune response. Meanwhile, research has shown that honey revealed to enhance the activity of the enzyme plasmin and plasmin destroys the fibrin which is committed to the debris in wound part, whereas it is not able to engulf the collagen matrix required for tissue repair (Frydman et al., 2020). By diminishing the concentration of prostaglandins and augmenting nitric oxide, honey may stimulate antibody synthesis and improve humoral immune responses. Additionally, the existence of prebiotic oligosaccharides in honey empowers it to assistance and exaggerates the immune system (Alshehabat et al., 2020). Nevertheless, honey-related exploration is truthfully promising to widen the conceivable therapeutic opportunities in wound restoration. Similarly, the persuasive biological efficacy of honey can promote new directions for improved choice of treatment approaches and signify a fascinating mechanism towards auspicious applications in wound healing (Figures 10.6 and 10.7).

FIGURE 10.6 Beneficial efficiency of honey in wound restoration mechanism.

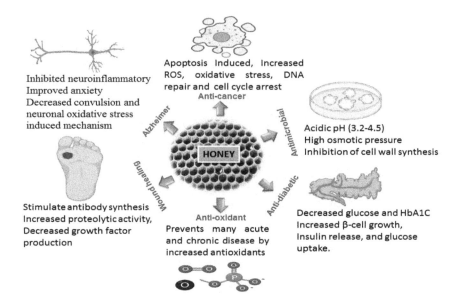

FIGURE 10.7   Summary of therapeutic and molecular mechanisms of anti-oxidative, anti-diabetic, antimicrobial, anticancer, neuroprotective and wound management effects of honey.

Conclusion

Pure honey can be an alternative natural agent with promising therapeutic characteristics in the medical scenario. In this chapter, we present scientific evidence of nutritional and health-enhancing effects of honey. Special attention has been paid to antioxidant, antimicrobial properties. Diabetes, cancer, Alzheimer's and wound healing are the most important ailments that honey can prevent. Further *in vitro* and *in vivo* scientific works are desirable to explore health-promoting effects and metabolic pathways of honey.

REFERENCES

Abubakar, M. B., Abdullah, W. Z., Sulaiman, S. A., Suen, A. B. 2012. A review of molecular mechanisms of the anti-leukemic effects of phenolic compounds in honey. *International Journal of Molecular Science* 13:15054–73.

Afrin, S., Haneefa, S. M., Fernandez-Cabezudo, M. J., Giampieri, F., Al-Ramadi, B. K., Battino, M. 2020. Therapeutic and preventive properties of honey and its bioactive compounds in cancer: an evidence-based review. *Nutrition Research Review* 33:50–76.

Ahmed, S., Othman, N. H. 2013. Honey as a potential natural anticancer agent: a review of its mechanisms. *Evidence Based Complementary and Alternative Medicine* 2013:829070.

Akanmu, M. A., Olowookere, T. A., Atunwa, S. A., Ibrahim, B. O., Lamidi, O. F., Adams, P. A., Ajimuda, B. O., Adeyemo, L. E. 2011. Neuropharmacological effects of

Nigerian honey in mice. *African Journal of Traditional, Complement Alternative Medicines* 8:230–49.

Al-Hatamleh, M. A. I., Boer, J. C., Wilson, K. L., Plebanski, M., Mohamud, R., Mustafa, M. Z. 2020. Antioxidant-based medicinal properties of stingless bee products: recent progress and future directions. *Biomolecules* 10:923.

Al-Koshab, M., Alabsi, A. M., Bakri, M. M., Naicker, M. S., Seyedan, A. 2020. Chemopreventive activity of Tualang honey against oral squamous cell carcinoma-in vivo. *Oral Surgery, Oral Medicine, Oral Pathology, Oral Radiology* 129:484–92.

Al-Rahbi, B., Zakaria, R., Othman, Z., Hassan, A., Ahmad, A. H. 2014. The effects of Tualang honey supplement on medial prefrontal cortex morphology and cholinergic system in stressed ovariectomised rats. *International Journal of Applied Research in Natural Products* 7:28–36.

Alshehabat, M., Hananeh, W., Ismail, Z. B., Rmilah, S. A., Abeeleh, M. A. 2020. Wound healing in immunocompromised dogs: A comparison between the healing effects of moist exposed burn ointment and honey. *Veterinary World* 13:2793–7.

Arora, R. S., Arora, P. R., Seth, R., Sharma, S., Kumar, C., Dhamankar, V., Kurkure, P., Prasad, M. 2020. Childhood cancer survivorship and late effects: The landscape in India in 2020. *Pediatric Blood and Cancer* 67:28556.

Aziz, C. B. A., Nazariah Ismail, C. A., Hussin, C. M. C., Mohamed, M. 2014. The antinociceptive effects of Tualang Honey in male Sprague-Dawley rats: A preliminary study. *Journal of Traditional Complementary Medicine* 4:298–302.

Aziz, M. S. A., Giribabu, N., Rao, P. V., Salleh, N. 2017. Pancreatoprotective effects of *Geniotrigonathoracica* stingless bee honey in streptozotocin-nicotinamide-induced male diabetic rats. *Biomedicine and Pharmacotherapy* 89:135–45.

Azman, K. F., Zakaria, R., Abdul Aziz, C. B., Othman, Z., Al-Rahbi, B. 2015. Tualang honey improves memory performance and decreases depressive-like behavior in rats exposed to loud noise stress. *Noise and Health* 17:83–9.

Azman, K. F., Zakaria, R., Abdul Aziz, C. B., Othman, Z. 2016. Tualang honey attenuates noise stress-induced memory deficits in aged rats. *Oxidative Medicine and Cellular Longevity* 2016:1549158.

Badolato, M., Carullo, G., Cione, E., Aiello, F., Caroleo, M. C. 2017. From the hive: Honey, a novel weapon against cancer. *European Journal of Medicinal Chemistry* 142:290–9.

Bang, L. M., Buntting, C., Molan, P. 2003. The effect of dilution on the rate of hydrogen peroxide production in honey and its implications for wound healing. *Journal of Alternative and Complementary Medicine* 9:267–73.

Baranowska-Wojcik, E., Szwajgier, D., Winiarska-Mieczan, A. 2020. Honey as the potential natural source of cholinesterase inhibitors in Alzheimer's disease. *Plant Foods for Human Nutrition* 75:30–2.

Batumalaie, K., Qvist, R., Yusof, K. M., Ismail, I. S., Sekaran, S. D. 2014. The antioxidant effect of the Malaysian Gelam honey on pancreatic hamster cells cultured under hyperglycemic conditions. *Clinical and Experimental Medicine* 14:185–95.

Batumalaie, K., Safi, S. Z., Yusof, K. M., Bin Ismail, I. S., Sekaran, S. D., Qvist, R. 2013. Effect of Gelam honey on the oxidative stress induced signaling pathways in pancreatic hamster cells. *International Journal of Endocrinology* 2013:1–10.

Blackman, L. D., Oo, Z. Y., Qu, Y., Gunatillake, P. A., Cass, P., Locock, K. E. S. 2020. Antimicrobial honey-inspired glucose-responsive nanoreactors by polymerization-induced self-assembly. *ACS Applied Materials and Interfaces* 12:11353–62.

Bobis, O., Moise, A. R., Ballesteros, I., Reyes, E. S., Durán, S. S., Sanchez-Sanchez, J., Cruz-Quintana, S., Giampieri, F., Battino, M., Alvarez-Suarez, J. M. 2020. Eucalyptus honey: Quality parameters, chemical composition and health-promoting properties. *Food Chemistry* 325:126870.

Bouzo, D., Cokcetin, N. N., Li, L., Ballerin, G., Bottomley, A. L., Lazenby, J., Whitchurch, C. B., Paulsen, I. T., Hassan, K. A., Harry, E. J. 2020. Characterizing the mechanism of action of an ancient antimicrobial, manuka honey, against pseudomonas aeruginosa using modern transcriptomics. *Systems* 5:106–20.

Candiracci, M., Piatti, E., Dominguez-Barragan, M., Garcia-Antras, D., Morgado, B., Ruano, D., Gutierrez, J. F., Parrado, J., Castano, A. 2012. Anti-inflammatory activity of a honey flavonoid extract on lipopolysaccharide-activated N13 microglial cells. *Journal of Agricultural and Food Chemistry* 60:12304–11.

Carrier, A. 2017. Metabolic syndrome and oxidative stress: A complex relationship. *Antioxidants and Redox Signaling* 26:429–31.

Chauhan, A. S., Prinja, S., Srinivasan, R., Rai, B., Malliga, J. S., Jyani, G., Gupta, N., Ghoshal, S. 2020. Cost effectiveness of strategies for cervical cancer prevention in India. *PLOS ONE* 15:0238291.

Combarros-Fuertes, P. M., Estevinho, L., Teixeira-Santos, R. G., Rodrigues, A., Pina-Vaz, C., Fresno, J. M., Tornadijo, M. E. 2020. Antibacterial action mechanisms of honey: Physiological effects of avocado, chestnut, and polyfloral honey upon *staphylococcus aureus* and *Escherichia coli*. *Molecules* 25:1252.

Dai, Y., Jin, R., Verpoorte, R., Lam, W., Cheng, Y. C., Xiao, Y., Xu, J., Zhang, L., Qin, X. M., Chen, S. 2020. Natural deep eutectic characteristics of honey improve the bioactivity and safety of traditional medicines. *Journal of Ethnopharmacology* 250:112460.

Daneshzad, E., Keshavarz, S. A., Qorbani, M., Larijani, B., Azadbakht, L. 2020. Dietary total antioxidant capacity and its association with sleep, stress, anxiety, and depression score: A cross-sectional study among diabetic women. *Clinical Nutrition ESPEN* 37:187–94.

Das, S., Sarmah, S., Lyndem, S., Singha Roy, A. 2020. An investigation into the identification of potential inhibitors of SARS-CoV-2 main protease using molecular docking study. *Journal of Biomolecular Structure and Dynamics* 13:1–11.

El-Haskoury, R, Al-Waili, N, El-Hilaly, J, Al-Waili, W, Lyoussi, B. 2019. Antioxidant, hypoglycemic, and hepatoprotective effect of aqueous and ethyl acetate extract of carob honey in streptozotocin-induced diabetic rats. *Veterinary World* 12:1916–23.

Erejuwa, O. O., Sulaiman, S. A., Wahab, M. S., Salam, S. K., Salleh, M. S., Gurtu, S. 2010a. Antioxidant protective effect of glibenclamide and metformin in combination with honey in pancreas of streptozotocin-induced diabetic rats. *International Journal of Molecular Sciences* 11:2056–66.

Erejuwa, O. O., Gurtu, S., Sulaiman, S. A., Ab Wahab, M. S., Sirajudeen, K. N., Salleh, M. S. 2010b. Hypoglycemic and antioxidant effects of honey supplementation in streptozotocin-induced diabetic rats. *International Journal for Vitamin and Nutrition Research* 80:74–82.

Farhana, L., Sarkar, S., Nangia-Makker, P., Yu, Y., Khosla, P., Levi, E., Azmi, A., Majumdar, A. P. N. 2020. Natural agents inhibit colon cancer cell proliferation and alter microbial diversity in mice. *PLOS ONE* 15:0229823.

Fasanmade, A. A., Alabi, O. T. 2008. Differential effect of honey on selected variables in alloxan-induced and fructose induced diabetic rats. *African Journal of Biomedical Research* 11:191–6.

Folli, F., Corradi, D., Fanti, P., Davalli, A., Paez, A., Giaccari, A., Perego, C., Muscogiuri, G. 2011. The role of oxidative stress in the pathogenesis of type 2 diabetes mellitus micro- and macrovascular complications: Avenues for a mechanistic-based therapeutic approach. *Current Diabetes Reviews* 7:313–24.

Frydman, G. H., Olaleye, D., Annamalai, D., Layne, K., Yang, I., Kaafarani, H. M. A., Fox, J. G. 2020. Manuka honey microneedles for enhanced wound healing and the prevention and/or treatment of Methicillin-resistant *Staphylococcus aureus* (MRSA) surgical site infection. *Scientific Reports* 10:13229.

Goslinski, M., Nowak, D., Kłębukowska, L. 2020. Antioxidant properties and antimicrobial activity of manuka honey versus Polish honeys. *Journal of Food Science and Technology* 57:1269–77.

Habryka, C., Socha, R., Juszczak, L. 2020. The effect of enriching honey with propolis on the antioxidant activity, sensory characteristics, and quality parameters. *Molecules* 25:1176.

Hailu, D., Belay, A. 2020. Melissopalynology and antioxidant properties used to differentiate *Schefflera abyssinica* and polyfloral honey. *PLOS ONE* 15:0240868.

Hbibi, A., Sikkou, K., Khedid, K., El Hamzaoui, S., Bouziane, A., Benazza, D. 2020. Antimicrobial activity of honey in periodontal disease: A systematic review. *Journal of Antimicrobial Chemotherapy* 75:807–26.

Huang, J., Rui, W., Wu, J., Ye, M., Huang, L., Chen, H. 2020. Strategies for determining the bioactive ingredients of honey-processed Astragalus by serum pharmacochemistry integrated with multivariate statistical analysis. *Journal of Separation Science* 43:2061–72.

Johnston, M., McBride, M., Dahiya, D., Owusu-Apenten, R., Nigam, P. S. 2018. Antibacterial activity of Manuka honey and its components: An overview. *AIMS Microbiology* 4:655–64.

Khosla, S., Farr, J. N., Tchkonia, T., Kirkland, J. L. 2020. The role of cellular senescence in ageing and endocrine disease. *Nature Reviews Endocrinology* 16:263–75.

Kocarnik, J. 2020. Cancer's global epidemiological transition and growth. *Lancet* 395:757–8.

Lumbers, M. 2020. New tools in wound care to support evidence-based best practice. *British Journal of Community Nursing* 25:26–9.

Martinotti, S., Bucekova, M., Majtan, J., Ranzato, E. 2019. Honey: An effective regenerative medicine product in wound management. *Current Medicinal Chemistry* 26:5230–40.

Martinotti, S., Pellavio, G., Patrone, M., Laforenza, U., Ranzato, E. 2020. Manuka honey induces apoptosis of epithelial cancer cells through Aquaporin-3 and calcium signaling. *Life (Basel)* 10:256.

Masoura, M., Passaretti, P., Overton, T. W., Lund, P. A., Gkatzionis, K. 2020. Use of a model to understand the synergies underlying the antibacterial mechanism of $H_{(2)}$, $O_{(2)}$-producing honeys. *Scientific Reports* 10:17692.

Mohd Sairazi, N. S., Sirajudeen, K. N. S., Asari, M. A., Mummedy, S., Muzaimi, M., Sulaiman, S. A. 2017. Effect of Tualang honey against KA-induced oxidative stress and neurodegeneration in the cortex of rats. *BMC Complementary Medicine and Therapies* 17:31.

Mohd Sairazi, N. S., Sirajudeen, K. N. S., Muzaimi, M., Mummedy, S., Asari, M. A., Sulaiman, S. A. 2018. Tualang honey reduced neuroinflammation and caspase-3 activity in rat brain after kainic acid-induced status epilepticus. *Evidence Based Complementary and Alternative Medicine* 2018:7287820.

Munstedt, K., Männle, H. 2020. Bee products and their role in cancer prevention and treatment. *Complementary Therapies in Medicine* 51:102390.

Nazir, L., Samad, F., Haroon, W., Kidwai, S. S., Siddiqi, S., Zehravi, M. 2014. Comparison of glycaemic response to honey and glucose in type 2 diabetes. *Journal of Pakistan Medical Association* 64:69–71.

Nizet, O., Camby, S., Nizet, J. L. 2020. Use of honey dressings in wound healing. *Revue Medicale de Liege* 75:797–801.

Othman, Z., Shafin, N., Zakaria, R., Hussain, N. N., Mohammad, W. M. Z. W. 2011. Improvement in immediate memory after 16 weeks of Tualang honey (agro mas) supplement in healthy postmenopausal women. *Menopause* 18:1219–24.

Othman, Z., Zakaria, R., Hussain, N. H. N., Hassan, A., Shafin, N., Al-Rahbi, B., Ahmad, A. H. 2015. Potential role of honey in learning and memory. *Medical Sciences* 3:3–15.

Przybylski, C., Bonnet, V. 2021. Discrimination of isomeric trisaccharides and their relative quantification in honeys using trapped ion mobility spectrometry. *Food Chemistry* 341:128182.

Qaid, E. Y., Zakaria, R., Yusof, N. A., Sulaiman, S., Shafin, N., Ahmad, A., Othman, Z., Rahb, I. B., Muthuraju, S. 2021. Tualang honey prevents neuronal damage in medial prefrontal cortex (mPFC) through enhancement of cholinergic system in male rats following exposure to normobaric hypoxia. *Bangladesh Journal of Medical Science* 20:122–9.

Rahimi, R., Nikfar, S., Larijani, B., Abdollahi, M. 2005. A review on the role of antioxidants in the management of diabetes and its complications. *Biomedicine & Pharmacotherapy* 59:365–73.

Rahman, Q. B., Iocca, O., Kufta, K., Shanti, R. M. 2020. Global burden of head and neck cancer. *Oral and Maxillofacial Surgery Clinics* 32:367–75.

Rawson, T. M., Moore, L. S. P., Zhu, N., Ranganathan, N., Skolimowska, K., Gilchrist, M., Satta, G., Cooke, G., Holmes, A. 2020. Bacterial and fungal coinfection in individuals with coronavirus: A rapid review to support COVID-19 antimicrobial prescribing. *Clinical Infectious Diseases* 71:2459–68.

Rittschof, C. C., Vekaria, H. J., Palmer, J. H., Sullivan, P. G. 2018. Brain mitochondrial bioenergetics change with rapid and prolonged shifts in aggression in the honey bee, *Apis mellifera*. *Journal of Experimental Biology* 221:176917.

Sadeghi, F., Akhlaghi, M., Salehi, S. 2020. Adverse effects of honey on low-density lipoprotein cholesterol and adiponectin concentrations in patients with type 2 diabetes: A randomized controlled cross-over trial. *Journal of Diabetes & Metabolic Disorders* 19:373–80.

Safi, S. Z., Batumalaie, K., Qvist, R., Yusof, K. M., Bin Ismail, I. S. 2016. Gelam honey attenuates the oxidative stress-induced inflammatory pathways in pancreatic hamster cells. *Evidence Based Complementary and Alternative Medicine* 2016:1–13.

Sahlan, M., Rahmawati, O., Pratami, D. K., Raffiudin, R., Mukti, R. R., Hermasyah, H. 2020. The effects of stingless bee (Tetragonulabiroi) honey on streptozotocin-induced diabetes mellitus in rats. *Saudi Journal of Biological Sciences* 27:2025–30.

Saralaya, S., Thomas, N. S. 2021. Bee wax and honey—A primer for OMFS. *Oral and Maxillofacial Surgery* 25:1–6.

Shambaugh, P., Worthington, V., Herbert, J. H. 1990. Differential effects of honey, sucrose, and fructose on blood sugar levels. *Journal of Manipulative & Physiological Therapeutics* 13:322–5.

Sinyorita, S., Ghosh, C. K., Chakrabarti, A., Auddy, B., Ghosh, R., Debnath, P. K. 2011. Effect of Ayurvedic mercury preparation Makaradhwaja on geriatric canine—A preliminary study. *Indian Journal of Experimental Biology* 49:534–9.

Syarifah-Noratiqah, S., Naina-Mohamed, I., Zulfarina, M. S., Qodriyah, H. M. 2018. Natural polyphenols in the treatment of Alzheimer's disease. *Current Drug Targets* 19:927–37.

Talebi, M., Farkhondeh, T., Samarghandian, S. 2020. Molecular mechanism-based therapeutic properties of honey. *Biomedicine & Pharmacotherapy* 130:110590.

Ullah, A., TlakGajger, I., Majoros, A., Dar, S. A., Khan, S., Mullah, K., Haleem Shah, A., Nasir Khabir, M., Hussain, R., Khan, H. U., Hameed, M., Anjum, S. I. 2021. Viral impacts on honey bee populations: A review. *Saudi Journal of Biological Sciences* 28:523–30.

Vogt, N. A., Vriezen, E., Nwosu, A., Sargeant, J. M. A. 2021. Scoping review of the evidence for the medicinal use of natural honey in animals. *Frontiers in Veterinary Science* 7:618301.

Wan Yaacob, W. M. H., Long, I., Zakaria, R., Zahiruddin, O. 2021. Tualang honey and its methanolic fraction ameliorate lipopolysaccharide-induced oxidative stress, amyloid deposition and neuronal loss of the rat hippocampus. *Advances in Traditional Medicine* 21:121–9.

Wang, X. Q., Wang, W., Peng, M., Zhang, X. Z. 2021a. Free radicals for cancer theranostics. *Biomaterials* 266:120474.

Wang, Y., Gou, X., Yue, T., Ren, R., Zhao, H., He, L., Liu, C., Cao, W. 2021b. Evaluation of physicochemical properties of Qinling *Apis cerana* honey and the antimicrobial activity of the extract against Salmonella Typhimurium LT(2) in vitro and in vivo. *Food Chemistry* 337:127774.

Yahaya, R., Mohd, N. Z., Zahiruddin, O., Rusli, I., Nik Ahmad, S. N. H., Aniza Abd, A., Rahima, D., Azizul, F. J. 2020. Tualang honey supplementation as cognitive enhancer in patients with schizophrenia. *Heliyon* 6:03948.

Yilmaz, A. C., Aygin, D. 2020. Honey dressing in wound treatment: A systematic review. *Complementary Therapies in Medicine* 51:102388.

Yu, L., Palafox-Rosas, R., Luna, B., She, R. C. 2020. The bactericidal activity and spore inhibition effect of Manuka honey against *Clostridioides difficile*. *Antibiotics (Basel).* 9:684.

Zubai, R. R., Aziz, N. 2015. As smooth as honey—The historical use of honey as topical medication. *JAMA Dermatology* 151:1102.

Index

For Product Safety Concerns and Information please contact our EU
representative GPSR@taylorandfrancis.com
Taylor & Francis Verlag GmbH, Kaufingerstraße 24, 80331 München, Germany

www.ingramcontent.com/pod-product-compliance
Ingram Content Group UK Ltd.
Pitfield, Milton Keynes, MK11 3LW, UK
UKHW021109180425
457613UK00001B/5